Jordan, Real and Lie Structures in Operator Algebras

Mathematics and Its Applications

Volume 418

Jordan, Real and Lie Structures in Operator Algebras

by

Shavkat Ayupov

Institute of Mathematics,
Academy of Sciences,
Tashkent, Uzbekistan

Abdugafur Rakhimov

University of World Economy and Diplomacy,
Tashkent, Uzbekistan

and

Shukhrat Usmanov

Institute of Mathematics,
Academy of Sciences,
Tashkent, Uzbekistan

Springer-Science+Business Media, B.V.

A C.I.P. Catalogue record for this book is available from the Library of Congress.

ISBN 978-90-481-4891-2 ISBN 978-94-015-8605-4 (eBook)
DOI 10.1007/978-94-015-8605-4

Printed on acid-free paper

Dedicated to the memory

of our teacher,

Professor T.A.Sarymsakov

CONTENTS

INTRODUCTION 1

PRELIMINARIES 7

Chapter 1: JORDAN OPERATOR ALGEBRAS 13

1.1. JW-algebras and enveloping von Neumann algebras 14

1.2. Traces on JW-algebras 30

1.3. Types of JW-algebras and enveloping von Neumann algebras 38

1.4. Classification of type I real and Jordan factors 44

1.5. Involutive antiautomorphisms of W^*-algebras 48

1.6. Involutive antiautomorphisms of injective W^*-algebras 58

1.7. Classification of injective real and Jordan factors type II_1, II_∞, and III_λ, $0 < \lambda \leq 1$ 66

COMMENTS TO CHAPTER 1 70

Chapter 2: REAL STRUCTURE IN W^*-ALGEBRAS 72

2.1. Real crossed products of real W^*-algebras by an automorphism 73

2.2. Discrete decomposition of real type III_0 factors 92

2.3. Periodic antiautomorphisms and automorphisms of complex and real type II_1 factors 110

2.4. Outer conjugacy classes of antiautomorphisms and automorphisms of complex and real factors 130

2.5. Injectivity, amenability, semidiscreteness and hyperfiniteness in real W^*-algebras 141

2.6. Diameters of state spaces of JW-algebras 151

COMMENTS TO CHAPTER 2 172

Chapter 3: LIE STRUCTURE IN OPERATOR ALGEBRAS 174

3.1. Theorem on isomophism of prime Lie rings. 175

3.2. Symmetric and skew-symmetric operators on real and quaternionian Hilbert spaces 182

3.3. Commutators of skew elements in real factors. 195

3.4. Isomorphism of Lie operator algebras and conjugacy of involutive antiautomorphisms. 206

COMMENTS TO CHAPTER 3 210

REFERENCES 212

INDEX 224

INTRODUCTION

The theory of operator algebras acting on a Hilbert space was initiated in thirties by papers of Murray and von Neumann. In these papers they have studied the structure of algebras which later were called von Neumann algebras or W^*-algebras. They are weakly closed complex *-algebras of operators on a Hilbert space. At present the theory of von Neumann algebras is a deeply developed theory with various applications.

In the framework of von Neumann algebras theory the study of factors (i.e. W^*-algebras with trivial centres) is very important, since they are comparatively simple and investigation of general W^*-algebras can be reduced to the case of factors. Therefore the theory of factors is one of the main tools in the structure theory of von Neumann algebras.

In the middle of sixtieth Topping [To 1] and Stormer [S 2] have initiated the study of Jordan (non associative and real) analogues of von Neumann algebras - so called JW-algebras, i.e. real linear spaces of self-adjoint operators on a complex Hilbert space, which contain the identity operator $\mathbf{1}$, closed with respect to the Jordan (i.e. symmetrised) product

$x \circ y = \frac{1}{2}(xy + yx)$ and closed in the weak operator topology.

The structure of these algebras has happened to be close to the structure of von Neumann algebras and it was possible to apply ideas and methods similar to von Neumann algebras theory in the study of JW-algebras. Thus Topping [To 1] has classified JW-algebras into those of type I, II_1, II_∞, III, later Stormer [S 2], [S 4] and Ayupov [A 1], [A 3], [A 4], [A 9], [A 10], [A 13] considered the problem on connections between the type of a JW-algebra and the type of its enveloping von Neumann algebra. In [S 2] Stormer gave a complete study of type I JW-algebras and has proved also that any reversible JW-algebra A (in particular of type II and III) is isomorphic to the direct sum $A_c \oplus A_r$, where the JW-algebra A_c is the self-adjoint part $\mathcal{U}(A_c)_s$ of its enveloping von Neumann algebra $\mathcal{U}(A_c)$, whence the JW-algebra A_r coincides with the self-adjoint part $R(A_r)_s$ of the real enveloping algebra $R(A_r)$ such that $R(A_r) \bigcap iR(A_r) = \{0\}$ (so called real von Neumann algebra). In this connection the study of real von Neumann algebras was carried out parallel to the theory of JW-algebras.

A real von Neumann algebra is a real *-algebra $\mathfrak{R}$ of bounded linear operators on a complex Hilbert space, containing the identity operator $\mathbf{1}$, which is closed in the weak operator topology and satisfies the condition $\mathfrak{R} \bigcap i\mathfrak{R} = \{0\}$. The smallest (complex) von Neumann algebra $\mathcal{U}(\mathfrak{R})$ containing $\mathfrak{R}$ coincides with its complexification $\mathfrak{R} + i\mathfrak{R}$, i.e., $\mathcal{U}(\mathfrak{R}) = \mathfrak{R} + i\mathfrak{R}$. Moreover $\mathfrak{R}$ generates a natural involutive (i.e. of order 2) *-antiautomorphism $\alpha_{\mathfrak{R}}$ of $\mathcal{U}(\mathfrak{R})$, namely

$$\alpha_{\mathfrak{R}}(x + iy) = x^* + iy^*,$$

where $x + iy \in \mathcal{U}(\mathfrak{R})$, $x, y \in \mathfrak{R}$.

It is clear that $\mathfrak{R} = \{x \in \mathcal{U}(\mathfrak{R}) : \alpha_{\mathfrak{R}}(x) = x^*\}$. Conversely, given a von Neumann algebra $\mathcal{U}$ and any involutive *-antiautomorphism α on $\mathcal{U}$, the set $\{x \in \mathcal{U} : \alpha(x) = x^*\}$ is real von Neumann algebra.

It is not difficult to see that two real von Neumann algebras generating the same von Neumann algebra, are isomorphic if and only if the corresponding involutive *-antiautomorphisms are conjugate [A 3], [A 6], [S 3], [S 4].

The theory of real von Neumann algebras has been developed intensively recent years. Starting with results of Giordano [G 3],[G 4] on classification of involutive *-antiautomorphisms of injective factors, Ayupov has defined and classified injective JW-factors and injective real factors [A 5], [A 9]. The structure of type II_∞ real von Neumann algebras was investigated by Stacey [St 1] and of real von Neumann algebras of type III - by Stacey [St 4] and independently by Usmanov [U 1], [U 2], [U 3], [U 4].

Thus the study of real von Neumann algebras and JW-algebras can be reduced to the study of pairs $(\mathcal{U}, \alpha)$, where $\mathcal{U}$ is a von Neumann algebra, α - its involutive *-antiautomorphism . In this case the set

$$\Re_\alpha = \{x \in \mathcal{U} : \alpha(x) = x^*\}$$

forms a real von Neumann algebra, while the set

$$A = \{x \in \mathcal{U} : \alpha(x) = x = x^*\} = \Re_s^\alpha$$

is a JW-algebra. This JW-algebra can be also considered as the self-adjoint part of the complex Jordan algebra

$$\mathcal{U}^\alpha(+1) = \{x \in \mathcal{U} : \alpha(x) = x\}.$$

Therefore it is natural to consider one more structure connected with the involutive *-antiautomorphism α, namely,

$$\mathcal{U}^\alpha(-1) = \{x \in \mathcal{U} : \alpha(x) = -x\}.$$

which is a (complex) Lie algebra with respect to the brackets $[x, y] = xy - yx$, and its skew part

$$\mathcal{U}^\alpha(-1)_k = \{x \in \mathcal{U} : \alpha(x) = -x = x^*\}$$

is exactly the skew part of the real von Neumann algebra $\Re^\alpha$, i.e.,

$$\mathcal{U}^\alpha(-1)_k = \{x \in \Re^\alpha : x^* = -x\} = \Re_k^\alpha.$$

In this connection Robinson and Stormer [RS] have initiated the study of pairs of Jordan and Lie algebras $\mathcal{U}^\alpha(\pm 1)$, generated by an involutive *-antiautomorphism α of a von Neumann algebra $\mathcal{U}$. They have considered in details the case when $\mathcal{U}$ is a type I factor, i.e., $\mathcal{U} = B(H)$, where H is a complex Hilbert space. Lie algebras of the form $\mathcal{U}^\alpha(-1)$

were investigated also in the monograph [Har 1] of de la Harpe.

Lie algebras $\mathcal{U}^\alpha(-1)$ and $\mathcal{U}^\alpha(-1)_k$ for arbitrary factors $\mathcal{U}$ were considered by Ayupov [A 15], [A 16], [A 17], [A 18], Ayupov and Azamov [AAz], etc.

Therefore given an involutive *-antiautomorphism α of a von Neumann algebra $\mathcal{U}$ one can consider the following three algebraic structures in $\mathcal{U}$:

real :

$$\mathfrak{R}^\alpha = \{x \in \mathcal{U} : \alpha(x) = x^*\};$$

Jordan:

$$\mathcal{U}^\alpha(+1) = \{x \in \mathcal{U} : \alpha(x) = x\};$$

$$\mathcal{U}^\alpha(+1)_s = \mathfrak{R}^\alpha_s;$$

and **Lie**:

$$\mathcal{U}^\alpha(-1) = \{x \in \mathcal{U} : \alpha(x) = -x\},$$

$$\mathcal{U}^\alpha(-1)_k = \mathfrak{R}^\alpha_k.$$

The present monograph is devoted to the study of each of these algebraic structures in von Neumann algebras, especially in the case of factors. Relation between these structures are also considered.

For the sake of clearness the book is devided into three chapters. The first chapter is devoted to JW-algebras, in the second one we expose mainly recent results on real von Neumann algebras, in the third chapter we study the Lie structure in von Neumann algebras. In comments we give an observation of references.

The first chapter and partially the second one can be considered as a further development of results exposed in the monograph [A 9] (see also [SAKC], [H-OS]).

In the first chapter we give the theory of JW-algebras and their type classification. Relations between the types of a JW-algebra and its enveloping von Neumann algebra are studied, normal traces on JW-algebras and their extension to von Neumann algebras are considered. We describe JW-factors of type I, and for continuous JW-factors we study their relation with involutive *-antiautomorphisms of enveloping von Neumann algebras. Besides we consider the conjugacy classes of involutive *-antiautomorphisms of injective factors, introduce and describe injective JW-factors of type II_1, II_∞ and III, give their construction.

The second chapter is devoted to real von Neumann algebras. We obtain the discrete decomposition of real type III factors, study periodic *-automorphisms and *-antiautomorphisms of real and complex hyperfinite II_1 - factors which play a basic role in the study of general hyperfinite factors. Outer conjugacy classes of *-antiautomorphisms for complex hyperfinite factors and of *-automorphisms for real hyperfinite factors are described. Finally the notion of injectivity, hyperfiniteness, amenability and semidiscreteness for real von Neumann algebras are studied which play an important role in the classical theory on injective factors. We obtain conditions for equivalence of these properties which differ from the case of complex von Neumann algebras.

The third chapter is devoted to the Lie structures in real and complex factors. Given a real factor $\mathfrak{R}$ we investigate whether the Lie algebra $\mathfrak{R}_k$ or its derived Lie algebra $[\mathfrak{R}_k, \mathfrak{R}_k]$ uniquely determine the associative structure of $\mathfrak{R}$. To this aim we obtain rather deep results concerning the structure of a single operator on a Hilbert space.

As a corollary we prove that given two involutive *-antiautomorphisms α and β of a complex factor $\mathcal{U}$, the complex Lie algebras $\mathcal{U}^\alpha(-1)$ and $\mathcal{U}^\beta(-1)$ are isomorphic if and only if α and β are conjugate, i.e., $\alpha = \theta^{-1}\beta\theta$ for an appropriate *-automorphism θ of $\mathcal{U}$. Similar results on isomorphism between real von Neumann algebras $\mathfrak{R}^\alpha$ and $\mathfrak{R}^\beta$ and between Jordan algebras $\mathcal{U}^\alpha(+1)$ and $\mathcal{U}^\beta(+1)$ we proved earlier and were comparatively less difficult.

We use the transparant numeration of theorems and propositions throughout the book: the first digit denotes the chapter, the second denotes the section and the third number denotes the number of the theorem or proposition.

In the beginning of the monograph we give some preliminaries from the theory of von Neumann algebras, which we hope are sufficient for understanding of the main exposition. For details of the theory of operator algebras we refer the reader to monographs [A 9], [JSSS], [Sak 1], [SAKC], [SZ], [T 1], [H-OS].

PRELIMINARIES

Let H be a complex Hilbert space, $B(H)$ denote the algebra of all bounded linear operators on H. The *weak (operator) topology* on $B(H)$ is the locally convex topology, generated by seminorms of the form:

$$\rho(a) = |(\xi, a\eta)|, \quad \xi, \eta \in H,\ a \in B(H).$$

W^* - a l g e b r a - is a weakly closed complex $*$-algebra of operators on a Hilbert space H containing the identity operator $\mathbf{I}$. Recall that W^*-algebras are also called *von Neumann algebra*.

We give another useful definition of von Neumann algebras:

a C^*-algebra M (i.e. the Banach *-algebra with properties $\|x\|^2 = \|xx^*\|$, for all $x \in M$) is called a von Neumann algebra if it is a dual space as a Banach space, i.e. there exists a Banach space M_* such that $(M_*)^* = M$, where $(M_*)^*$ is the dual Banach space of M_*. We shall call such a Banach space M_* the predual of M. We remind that M_* is isomorphic to the set of all normal linear functionals of M [Sak 1, 1.13].

Let M be a W^*-algebra. The set M' of all elements from $B(H)$ commuting with each element from M is called the *commutant* of the algebra M. The *center* $Z(M)$ of a W^*-algebra M is the set of elements of M, commuting with each element from M. It is easy to see that $Z(M) = M \bigcap M'$. Elements of $Z(M)$ are called *central* elements. A W^*-algebra M is called *factor*, if $Z(M)$ consists of the complex multiples of $\mathbf{I}$, i.e if $Z(M) = \{\lambda\mathbf{I},\ \lambda \in \mathbb{C}\}$.

Let e, f, h be projections from M. We say that e is equivalent to f, and write $e \sim f$, if $e = w^*w$, $f = ww^*$ for some partial isometry w from M.

A projection e is called:

finite, if $e \sim f \leq e$ implies $f = e$;

infinite - otherwise ;

purely infinite, if e doesn't have any nonzero finite subprojection ;

abelian, if the algebra eMe is an abelian W^*-algebra.

A W^*-algebra M is called

finite, infinite, purely infinite, if $\mathbf{I}$ is a finite, infinite, purely infinite respectively;

σ-finite, if any family of pairwise orthogonal projections from M is at most countable;

semifinite, if each projection in M contains a nonzero finite subprojection ;

properly infinite, if every nonzero projection from $Z(M)$ is infinite;

discrete, or of *type* I, if it contains a faithful abelian projection (i.e. an abelian projection with the central support $\mathbf{I}$);

continuous, if there is no abelian projection in M except zero ;

the *type* II, if M is semifinite and continuous;

the *type* I_{fin} (respectively I_∞), if M is of type I and finite (respectively properly infinite);

the *type* II_1 (respectively *type* II_∞), if M is of type II and finite (respectively properly infinite);

the *type* III, if M is purely infinite.

Theorem 1 [T 1]. *Any W^*-algebra has a unique decomposition along its center into the direct sum of W^*-algebras of the following types:*

1) the type I_{fin} ;

2) the type I_∞ ;

3) the type II_1 ;

4) the type II_∞ ;

5) the type III.

A linear mapping $\alpha : M \to M$ is called a **-automorphism* (respectively a *-antiautomorphism) if

1) $\alpha(x^*) = \alpha(x)^*$;

2) $\alpha(xy) = \alpha(x)\alpha(y)$ (respectively $\alpha(xy) = \alpha(y)\alpha(x)$),

for all $x, y \in M$. A *-automorphism α is called *inner* if there exists a unitary u in M, such that $\alpha(x) = Adu(x) = uxu^*$, for all $x \in M$. We shall denote by $Aut(M)$ the group of all *-automorphisms , by $Ant(M)$ the group of all *-antiautomorphisms, and by $Int(M)$ the group of all inner *-automorphisms of M.

Two *-automorphisms or *-antiautomorphisms α and β are said to be *conjugate* (or *outer conjugate*), if $\alpha = \theta \cdot \beta \cdot \theta^{-1}$ (respectively $Adu \cdot \alpha = \theta \cdot \beta \cdot \theta^{-1}$) for some *-automorphism θ (and an inner *-automorphism Adu).

A linear functional ω on M is called *positive*, if $\omega(x^*x) \geq 0$ for all $x \in M$. A positive linear functional ω on M with $\|\omega\| = 1$ is called a *state*. Let M_+ be the positive part of M. A *weight* on M is a homogeneous additive function $\omega : M_+ \to [0, +\infty]$ (we suppose that $0 \cdot +\infty = 0$).

A weight (or a state) ω is called:

faithful, if for any $x \in M_+$, $\omega(x) = 0$ implies $x = 0$;

normal, if for any net $\{x_\alpha\}$ in M, increasing to an element x, we have $\omega(x) = \sup_\alpha \omega(x_\alpha)$;

finite, if $\omega(x) < \infty$ for all $x \in M_+$. In this case $\omega \in M^*$; and if ω is normal also then $\omega \in M_*$;

semifinite, if for any $x \in M_+$ there exists a net of elements $\{y_\alpha\} \in M_+$, such that $\omega(y_\alpha) < \infty$, and $y_\alpha \to x$ in σ - weak topology ;

trace, if $\omega(uxu^*) = \omega(x)$ for all $x \in M_+$ and each unitary $u \in M$.

The type of a W^*-algebra is tightly connect with the existence of traces on it. Namely a W^*-algebra M is a finite if and only if it possesses a separating family of finite normal traces; it is semifinite if and only if it possesses a faithful semifinite normal trace; M is purely infinite if and only if there is no nonzero semifinite normal trace on M (see [T 1] or [Sak 1]).

Let ω be a faithful normal semifinite weight (such weight exists on every W^*-algebra M (see Proposition 2.7.13 [BR])). Put

$$L_\omega = \{x \in M : \omega(x^*x) < \infty\}.$$

This is a left ideal in M. Define in L_ω an inner product by $< x, y >= \omega(x^*y)$, for $x, y \in L_\omega$ and denote by $\mathcal{H}$ the completion of L_ω in the norm, generated by this inner product. The imbedding of L_ω into the Hilbert space $\mathcal{H}$ is denoted by γ. The left regular representation of M in $\mathcal{H}$ is the representation π on $\gamma(L_\omega)$ defined by

$$\pi(x)\gamma(y) = \gamma(xy), \text{ for } x \in M, y \in L_\omega.$$

Now we shall give a brief account of the main definition and results from the Tomita-Takesaki's theory. First we shall define in $\mathcal{H}$ the following two $\mathbf{R}$-linear ($\mathbf{C}$ - linear) operators S_0 and F_0: for $x \in L_\omega$, $x' \in L'_\omega$ (L'_ω is the commutant of L_ω in $\pi(M)$) put

$$S_0(\gamma(x)) = \gamma(x^*), \qquad F_0(\gamma(x')) = \gamma((x')^*).$$

Theorem 2 [BR]. *The antilinear operators S_0 and F_0 has closures in $\gamma(L_\omega)$ and $\gamma(L'_\omega)$ respectively , which are denoted by $S = \overline{S}_0$ and $F = \overline{F}_0$ respectively .*

If ω is a trace, then the operator S is an antiunitary, and $\|S(\xi)\|_{\mathcal{H}} = \|\xi\|_{\mathcal{H}}$ for all $\xi \in \mathcal{H}$. In this case the inner automorphism of the algebra $B(\mathcal{H})$, generated by $S : x \to SxS$ $(S^2 = \mathbf{1})$, maps M in M', and in this case $SMS = M'$. In general case S doesn't preserve the norm in $\mathcal{H}$ and differs from an antiunitary operator by some self-adjoint co-factor, which is denoted by $\Delta^{1/2}$: $S = J\Delta^{1/2}$, where J is an antiunitary operator, and Δ is a positive self-adjoint ($\mathbf{C}$ - linear) operator. The operator J is called *modular involution*, and Δ is called *modular operator.*

The Tomita-Takesaki's theory investigates the connection between S, F, J, Δ and the structure of M and M'.

Theorem 3 [BR]. *The following relation between S, F, J and Δ are valid:*

$$S^* = F, \qquad F^* = S,$$

$$SF = \Delta^{-1}, \qquad FS = \Delta,$$

$$J^2 = \mathbf{1}, \qquad \Delta^{-1/2} = J\Delta^{-1/2}J,$$

$$F = \Delta^{1/2}J, \qquad S = J\Delta^{1/2}.$$

If ω is a trace, then $\Delta = \mathbf{1}$, $S = J$.

The following theorem is the main result of Tomita-Takesaki's theory.

Theorem 4 [BR]. *For any faithful normal semifinite weight one has*

$$JMJ = M', \qquad \Delta^{it}M\Delta^{-it} = M,$$

for all $t \in \mathbf{R}$.

Thus M is antiisomorphic to M', and $\Delta = \Delta_\omega$ generates a *-automorphisms group in M, which is denoted by $\{\sigma_t^\omega,\ t \in \mathbf{R}\}$.

Let M be a factor; put $S(M) = \bigcap_\omega Spec\,\Delta_\omega$, where ω run over the set all normal semifinite weights on M, and $Spec\,\Delta_\omega$ is the spectrum of the self-adjoint modular operator Δ_ω, associated with ω.

Theorem 5 [C 1]. *$S(M)\setminus\{0\}$ is a multiplicative subgroup of $\mathbf{R}_+$ and $S(M)$ may be equal to only one of the following four sets:*

1) $\{1,0\}$;

2) $[0,\infty)$;

3) $\{0\}\cup\{\lambda^n\}$, $\lambda\in(0,1), n\in Z$;

4) $\{1\}$.

Definition 6 [C 1]. We say that a factor M has the *type III_1* if $S(M)=[0,\infty)$; the *type III_λ*, if $S(M)=\{0\}+\{\lambda^Z\}, 0<\lambda<1$ and the *type III_0*, if $S(M)=\{0,1\}$.

All other details and results we refer the reader to [A 9], [A 17], [A 16], [BR], [G 2], [C 1], [C 4], [T 1], [To 1], [Sak 1], [H-OS].

CHAPTER 1

JORDAN OPERATOR ALGEBRAS

This chapter is devoted to weakly closed Jordan algebras of self-adjoint operators on a complex Hilbert space - so called *JW*-algebras introduced by D.Topping. They present a real and non associative counterpart for von Neumann algebras, and many parts of the structure theory of *JW*-algebras are very similar to the corresponding parts of von Neumann algebras theory.

In Section 1.1 we define the notion of modular projection which is the analogue of finite projection in W^*-algebras, classify *JW*-algebras into modular, locally modular, properly non modular and purely non modular types. We study the structure of the enveloping von Neumann algebra for a *JW*-algebra, its relations with the given *JW*-algebra and thus prepare a very useful instrument for investigation of Jordan operator algebras.

Section 1.2 is devoted to traces on *JW*-algebras and the problem of their extension to the enveloping W^*-algebra. We also obtain type criterions for *JW*-algebras and real W^*-algebras in terms of the existence of normal traces.

In Section 1.3 we prove that the type of a reversible *JW*-algebra coincides with the type of its enveloping von Neumann algebra.

In Section 1.4 we describe type I JW-factors and show that up to isomorphism they are exactly Jordan algebras of self-adjoint operators on real, complex or quaternionian Hilbert space or spin factors. We also give a description of real W^*-factors of type I.

In Section 1.5 we show that JW-factors non isomorphic to the self-adjoint part of a W^*-algebra are defined by involutive (i.e. of order 2) *-antiautomorphisms of enveloping W^*-algebras and the study of such JW-factors (up to isomorphism) can be reduced to the study of involutive *-antiautomorphisms of W^*-factors (up to conjugacy). This enables us to construct examples of JW-factors of any type which are not self-adjoint parts of von Neumann algebras.

Section 1.6 is devoted mainly to brief exposition of results of Giordano on classification up to conjugacy of involutive *-antiautomorphisms of injective factors. As corollaries in Section 1.7 we obtain classification (up to isomorphism) of injective JW-factors and real W^*-factors.

1.1. JW-algebras and enveloping von Neumann algebras

Let $B(H)$ be the algebra of all bounded linear operators on a complex Hilbert space H. We recall from [H-OS] that a *JW-algebra* is a real linear space of self-adjoint operators from $B(H)$ which is closed under the Jordan product $a \circ b = \frac{1}{2}(ab + ba)$ and also closed in the weak operator topology. A JW-algebra A is said to be *reversible* if $a_1 a_2 ... a_n + a_n a_{n-1} ... a_1 \in A$ whenever $a_1, a_2, ..., a_n \in A$.

Examples of non-reversible JW-algebras are given by spin factors which are defined as follows.

Example 1.1.1. A spin system in $B(H)$ is a set S of symmetries (i.e. self-adjoint unitaries) $\neq \pm \mathbf{1}$ in $B(H)$ which satisfy $s \circ t = st + ts = 0$

for all $s, t \in S$, $s \neq t$. It is obvious that if $(S)^-$ is the weak closure of the linear span of S, then any operator from $(S)^-$ is a multiple of some symmetry from $B(H)$. This implies that the linear space $\mathbf{R}\mathbf{1} + (S)^-$ is a JW-algebra. These JW-algebras are called *spin factors.*

In general, we can give an abstract definition of spin factor. Let A be a real Hilbert space with dimension ≥ 3 and let $< x|y >$ be the inner product in A, $x, y \in A$. Suppose that u is a unit vector in A, and $N = \{u\}^\perp$, i.e. $A = \mathbf{R}u + N$. Define a multiplication in A in the following way

$$(\alpha u + x) \circ (\beta u + y) = (\alpha\beta + < x|y >)u + (\beta x + \alpha y),$$

where $\alpha, \beta \in \mathbf{R}$, $x, y \in N$. Then A is a Jordan algebra with the identity element u, and it is called *abstract spin factor.* It is known that any abstract spin factor is isomorphic to some (concrete) spin factor and conversely any spin factor presents an example of abstract spin factors. Therefore, below abstract spin factors we will be also called spin factors. It is clear that the Jordan structure of a spin factor A is completely determined by the Hilbert space structure on A. Since Hilbert spaces are isomorphic if and only if they have the same dimension, it follows that two spin factors are isomorphic if and only if they have equal dimension, i.e. spin factor a completely determined by their dimensions (cf. [To 2]).

A real $*$-algebra $\mathfrak{R}$ in $B(H)$ is called a *real* W^**-algebra* if it is closed in the weak operator topology and satisfies the conditions $\mathfrak{R} \bigcap i\mathfrak{R} = \{0\}$, $\mathbf{1} \in \mathfrak{R}$. It is obvious that if R is a real or complex W^*-algebra, then its self-adjoint part $R_s = \{x \in R : x^* = x\}$ forms a reversible JW-algebra.

Given an arbitrary JW-algebra A let $\mathfrak{R}(A)$ denote the weakly closed real $*$-algebra in $B(H)$ generated by A, and let $\mathcal{U}(A)$ denote the W^*-algebra (complex) generated by A.

Proposition 1.1.2 [H-OS]. *Let A be a JW-algebra. Then there exist three orthogonal projections e, f, g in A with $e + f + g = \mathbf{1}$ such that*

1) eA is the self-adjoint part of the W^*-algebra $\mathcal{U}(eA)$;

2) fA is the self-adjoint part of the real W^*-algebra $\mathfrak{R}(fA)$, i.e. $\mathfrak{R}(fA) \bigcap i\mathfrak{R}(fA) = \{0\}$; in this case JW-algebra fA is called purely real *JW-algebra.*

3) gA is isomorphic to the direct sum of JW-algebras of the form $L^\infty(\Omega, \mu, V)$, where μ is a Radon measure on a locally compact space Ω, and *V is a spin factor. If, moreover, A is reversible then $dim(V) \leq 6$.*

Lemma 1.1.3. *Let $\Re$ be a uniformly closed real self-adjoint operator algebra with identity acting on a Hilbert space. Then the following three conditions are equivalent.*

(i) $\Re \bigcap i\Re = \{0\}$,

(ii) $(i\Re)^+ = \{0\}$,

(iii) *For all* a, b *in* $\Re$, $||a + ib|| \geq \max\{||a||, ||b||\}$.

In particular if the above conditions are satisfied then $\Re + i\Re$ is a C^*-algebra.

Proof. (i) $\Rightarrow$ (ii). If $a \in \Re$ and $b = ia \geq 0$ then $b^2 = a^*a$ is a positive operator in $\Re$. Since $\Re$ is self-adjoint and uniformly closed b, being the unique positive square root of b^2, belongs to $\Re$, hence to $\Re \bigcap i\Re = \{0\}$, $b = 0$.

(ii) $\Rightarrow$ (i). If $\Re \bigcap i\Re \neq \{0\}$ it is a C^*-algebra. Since $(\Re \bigcap i\Re)^* \subset (i\Re)^+ = \{0\}$, $\Re \bigcap i\Re = \{0\}$.

(ii) $\Rightarrow$ (iii). If $a + ib \geq 0$, $a, b \in \Re$ then $a \geq 0$. In fact, from the equivalence of (i) and (ii) a is self-adjoint, hence $a = a^+ - a^-$, $a^+, a^- \geq 0$ $a^+a^- = 0$. Let f be any continuous real function which is zero on the positive reals. Then

$$f(a)af(a) = -f(a)a^- f(a) \leq 0.$$

Since $f(a)(a + ib)f(a) \geq 0$, $f(a)ibf(a) \geq 0$. By (ii) $f(a)bf(a) = 0$, hence $f(a)af(a) = 0$. Since this holds for all such f, $a^- = 0$, i.e., $a \geq 0$. Let $a, b \in \Re$. In order to show $||a+ib|| \geq \max\{||a||, ||b||\}$, we may assume $||a + ib|| \leq 1$. Then

$$0 \leq \mathbf{1} - (a + ib)^*(a + ib) = \mathbf{1} - a^*a - b^*b - i(a^*b - b^*a).$$

By the above, $\mathbf{1} \geq a^*a + b^*b \geq 0$, so $\|a\| \leq 1$, $\|b\| \leq 1$, and (iii) follows. Clearly (iii) $\Rightarrow$ (i).

If the above conditions obtain let $\{c_n = a_n + ib_n\}$ be a Cauchy sequence in $\mathfrak{R} + i\mathfrak{R}$. Then $\|a_n - a_m + i(b_n - b_m)\| \to 0$ as $n, m \to \infty$. By (iii) $\|a_n - a_m\| \to 0$, $\|b_n - b_m\| \to 0$ as $n, m \to \infty$. Thus $\{a_n\}$ and $\{b_n\}$ are Cauchy sequences in $\mathfrak{R}$, hence converge to a and b in $\mathfrak{R}$ respectively. Thus $c_n = a_n + ib_n \to a + ib$ in $\mathfrak{R} + i\mathfrak{R}$, which is thus uniformly closed, hence is a C^*-algebra. ∎

As an immediate consequence of the proof of (ii) $\Rightarrow$ (iii) we have

Corollary 1.1.4. *Let $\mathfrak{R}$ be as in Lemma 1.1.3 and such that $\mathfrak{R} \bigcap i\mathfrak{R} = \{0\}$. If $a, b \in \mathfrak{R}$ are such that $a + ib \geq 0$, then $a \geq 0$.*

Theorem 1.1.5. *Let A be a reversible JW-algebra. Then $\mathcal{U}(A) = \mathfrak{R}(A) + i\mathfrak{R}(A)$.*

Proof. By Proposition 1.1.2 there exist central projections e and f in A such that eA is the self-adjoint part of a W^*-algebra, and $\mathfrak{R}(fA) \bigcap i\mathfrak{R}(fA) = \{0\}$, $e + f = \mathbf{1}$. Since the theorem holds for self-adjoint parts of W^*-algebras, we may assume $\mathbf{R}(A) \bigcap i\mathfrak{R}(A) = \{0\}$.

Let $B = \mathfrak{R}(A) + i\mathfrak{R}(A)$. Then $B \subset \mathcal{U}(A)$ and by Lemma 1.1.3 B is a C^*-algebra. In order to show $B = \mathcal{U}(A)$ it suffices to show B is a W^*-algebra. In order to accomplish this it suffices to show that the strong limit of each bounded monotone-increasing directed sequence of self-adjoint operators in B lies in B. Let a_α, b_α be operators in $\mathfrak{R}(A)$ such that $\{a_\alpha + ib_\alpha\}$ is a monotone-increasing directed sequence with least upper bound s. Then $a_\alpha + b_\alpha \to s$ strongly. The directed sequence $\{a_\alpha\}$ is monotone-increasing. Indeed, if $\alpha \geq \beta$ then $a_\alpha + ib_\alpha \geq a_\beta + ib_\beta$, so that $a_\alpha - a_\beta + i(b_\alpha - b_\beta) \geq 0$. By Corollary 1.1.4 $a_\alpha \geq a_\beta$. Moreover, by Lemma 1.1.3 $\|a_\alpha\| \leq \|s\|$. Let a be the least upper bound of the a_α. Then $a_\alpha \to a$ strongly, and $a \in \mathfrak{R}(A)$.

Let ξ be an arbitrary fixed vector, and let $\varepsilon > 0$. Since the directed sequences $\{a_\alpha\}$ and $\{a_\alpha + ib_\beta\}$ are monotone-increasing there exist α_0 and β_0 such that

$$\omega(a_\alpha) \leq \omega(a) \leq \omega(a_\alpha) + \varepsilon/2,$$

$$\omega(a_\beta + ib_\beta) \leq \omega(s) \leq \omega(\alpha_\beta + ib_\beta) + \varepsilon/2$$

for all $\alpha \geq \alpha_0, \beta \geq \beta_0$, where $\omega(a) = \omega_\xi(a) =< a\xi, \xi >$. Let $\gamma \geq \alpha_0, \beta_0$. Then

$$\begin{aligned} |\omega(ib_\gamma - (s-a))| &= |\omega(ib_\gamma) - \omega(s) + \omega(a_\gamma + ib_\gamma) - \omega(a_\gamma + ib_\gamma) + \omega(a)| \\ &\leq |-\omega(s) + \omega(a_\gamma + ib_\gamma)| + |\omega(a) - \omega(a_\gamma)| \\ &\leq \varepsilon/2 + \varepsilon/2 = \varepsilon. \end{aligned}$$

Thus $ib_\alpha \to s - a$ weakly, or, if $b = -i(s-a)$, then $b_\alpha \to b$ weakly. In particular $b \in \Re(A)$. Since also

$$|\omega(a_\gamma + ib_\gamma - (a + ib))| \leq |\omega(a_\gamma - a)|$$

$$+|\omega(ib_\gamma - (s-a))| \leq \varepsilon/2 + \varepsilon = 3\varepsilon/2,$$

$a_\alpha + ib_\alpha \to a + ib$ weakly. The weak topology is Hausdorff, and $\{a_\alpha + ib_\alpha\}$ converges to both $a + ib$ and s weakly. Thus

$$s = a + ib \in \Re(A) + i\Re(A) = B. \qquad \blacksquare$$

Further we shall describe the structure of the enveloping W^*-algebra for reversible JW-algebras. If a JW-algebra A is *abelian* (i.e. all its elements commute under the ordinary operator multiplication) then it coincides with the self-adjoint part of the abelian W^*-algebra $A + iA$ and therefore the problem is trivial. Thus without loss of generality we shall assume from now on that all considered JW-algebras *don't contain central abelian projections.*

Let A be a reversible JW-algebra, $\mathcal{U} = \mathcal{U}(A)$ its enveloping W^*-algebra.

Proposition 1.1.6. *If the JW-algebra A is isomorphic to the self-adjoint part of a W^*-algebra N then there are two central projections $e, f \in N$ with $e \vee f = \mathbf{1}$, such that $\mathcal{U}$ is *-isomorphic to $eN \oplus fN_0$,* *where N_0 is the opposite W^*-algebra for N.*

Proof. Let φ be an isomorphism between the JW-algebras N_s and A. For $x = a+ib \in N$, $a,b \in N_s$ put $\varphi(x) = \varphi(a)+i\varphi(b)$. Then we obtain a C^*-isomorphism φ between N and $A+iA \subset \mathcal{U}$. Since $\varphi(N) = A+iA$ generates $\mathcal{U}$ as a W^*-algebra, one can find by Lemma 3.2 from [S 1] a central projection $p \in \mathcal{U}$ such that the map $\varphi_1 : x \to \varphi(x)p$ is a *-homomorphism and the map $\varphi_2 : x \to \varphi(x)p^\perp$ is a *-anti-homomorphism from N to $\mathcal{U}$. The normality of φ (because φ is the isomorphism of JW-algebras) implies the normality of φ_1 and φ_2. Therefore, the kernels $Ker\varphi_1$ and $Ker\varphi_2$ are weakly closed two sided ideals in N, and hence $Ker\varphi_1 = q_1N$, $Ker\varphi_2 = q_2N$ for appropriate central projections q_1 and q_2 from N. Since $\varphi_1+\varphi_2 = \varphi$ and φ is faithful, the projections q_1 and q_2 are disjoint. It is clear that the quotient *-homomorphism $\psi_1 : N/_{Ker\varphi_1} \to (A+iA)p$ is a *-isomorphism and hence

$$N/_{Ker\varphi_1} = N/_{q_1N} \cong q_1^\perp N \cong (A+iA)p.$$

Therefore putting $e = q_1^\perp$ we obtain a *-isomorphism between the W^*-algebra eN and $(A+iA)p \subset \mathcal{U}p = \mathcal{U}(Ap)$, where $\mathcal{U}(Ap)$ is the enveloping W^*-algebra for the JW-algebra Ap. This implies that $(A+iA)p$ is a W^*-algebra containing Ap and therefore coincides with $\mathcal{U}(Ap)$. Thus we obtain the *-isomorphism

$$eN \cong (A+iA)p = \mathcal{U}(Ap) = \mathcal{U}(A)p.$$

Similarly if $f = q_2^\perp$ then fN and $\mathcal{U}p^\perp$ are *-anti-isomorphic and we obtain the *-isomorphism $fN_0 \cong \mathcal{U}p^\perp$. Therefore

$$\mathcal{U} = \mathcal{U}p \oplus \mathcal{U}p^\perp \cong eN \oplus fN_0,$$

where $e \vee f = q_1^\perp \vee q_2^\perp = (q_1q_2)^\perp = 0^\perp = \mathbf{1}$. ∎

Remark 1.1.7. From now on we identify *-isomorphic W^*-algebras $\mathcal{U}(A)$ and $eN \oplus fN_0 = \{(ea, fb_0), a, b \in N\}$. Then the proof of Proposition 1.1.6 shows that the JW-algebra A will coincide with the JW-algebra $\{(ea, fa_0), a \in N_s\}$, where a_0 denotes the element $a \in N$ considered as an element of the opposite W^*-algebra N_0. By the assumption the JW-algebra A and hence the W^*-algebra N do not contain central abelian projections, so by Lemma 10 from [S 1] one has $\Re(N_s) = N$, where $\Re(N_s)$ is the real *-algebra generated by the JW-algebra N_s. Therefore in this case

$$\Re(A) = \{(ea, fa_0^*),\ a \in N\}.$$

Proposition 1.1.8. *Let A be a purely real JW-algebra isomorphic to the self-adjoint part of a W^*-algebra N. Then $\mathcal{U}$ is *-isomorphic to the W^*-algebra $N \oplus N_0$.*

Proof. In view of Proposition 1.1.6 it is sufficient to show that $e = f = \mathbf{1}$. Consider the element $x = e - ef \in N_s$. Since $ex = x$ and $fx = 0$, one has from Remark 1.1.7

$$(x, 0) = (ex, fx_0) \in A \subset \Re(A).$$

At the same time $ix \in N$ and therefore from Remark 1.1.7 we have that

$$(ix, 0) = (e(ix), f(ix)_0^*) \in \Re(A)$$

i.e.

$$i(x, 0) = (ix, 0) = \in \Re(A) \bigcap i\Re(A) = \{0\}.$$

Thus $x = 0$, i.e. $e \leq f$. Similarly considering $y = f - ef$ we obtain $f \leq e$, i.e. $e = f$. From $e \vee f = \mathbf{1}$ it follows now that $e = f = \mathbf{1}$. ∎

Corollary 1.1.9. *Under the conditions of Proposition 1.1.6:*

(a) the JW-algebra A is purely real if and only if $e = f = \mathbf{1}$;

(b) the JW-algebra A coincides with the self-adjoint part of the W^*-algebra $\mathcal{U}(A)$ if and only if $ef = 0$.

Proof. (a) The necessity of the condition $e = f = \mathbf{1}$ is shown in Proposition 1.1.8. Conversely, let $e = f = \mathbf{1}$. Then

$$A = \{(a, a_0), a \in N_s\}, \qquad \Re(A) = \{(a, a_0^*), a \in N\}.$$

If $(a, a_0^*) = i(b, b_0^*) \in \Re(A) \bigcap i\Re(A)$ then $(ib)^* = ib^*$ and therefore $b = -b$, i.e. $b = 0$. Thus $\Re(A) \bigcap i\Re(A) = \{0\}$, i.e. A is purely real.

(b) If $ef = 0$ then from $e \vee f = \mathbf{1}$ it follows that $e^{\perp} = f$ and hence

$$\mathcal{U}(A) = eN \oplus e^{\perp} N_0.$$

If $(ea, e^{\perp} b_0) \in \mathcal{U}(A)_s$ then for $x = ea + e^{\perp} b \in N$ we have from Remark 1.1.7

$$(ea, e^{\perp} b_0) = (ex, e^{\perp} x_0) \in A.$$

Therefore $\mathcal{U}(A)_s \subset A$, i.e. $A = \mathcal{U}(A)_s$.

Conversely, if we suppose that $g = ef \neq 0$ then A contains the direct summand

$$A_0 = \{(ga, ga_0),\ a \in N_s\} = \{(b, b_0),\ b \in (gN)_s\},$$

which is a purely real JW-algebra in view of (a). This contradicts the condition $A = \mathcal{U}(A)_s$ which means that $\Re(A) = \mathcal{U}(A)$. ■

Corollary 1.1.10. *If a purely real JW-algebra A is isomorphic to the self-adjoint part of a W^*-algebra N, then $Z_A \subset_{\neq} (Z_{\mathcal{U}})_s$, where Z_A is the center of the JW-algebra A, $(Z_{\mathcal{U}})$ is the center of the W^*-algebra $\mathcal{U} = \mathcal{U}(A)$.*

Proof. From Remark 1.1.7 and Proposition 1.1.8 we have that $\mathcal{U} = N \oplus N_0$ and therefore

$$Z_{\mathcal{U}} = Z_N \oplus Z_N = \{(a, b_0),\ a, b \in Z_N\}.$$

On the other hand $Z_A = \{(a, a_0),\ a \in (Z_N)_s\}$.

It is clear that $Z_A \subset (Z_{\mathcal{U}})_s$ and for $a \in (Z_N)_s$ we have $(a, 0) \in (Z_{\mathcal{U}})_s \setminus Z_A$. ■

Now let A be a purely real JW-algebra, and let $\mathcal{U}(A) = \Re(A) + i\Re(A)$ be the enveloping W^*-algebra for A. Then any $x \in \mathcal{U}(A)$ can be uniquely represented in the form $x = a + ib$, $a, b, \in \Re(A)$. Put

$$\alpha(x) = a^* + ib^*.$$

Then α is an involutive (i.e. the map of order 2) *-antiautomorphism of the W^*-algebra $\mathcal{U}(A)$ and in particular α is a ultraweakly continuous.

Indeed, the injectivity of map α is trivial. From $a + ib = \alpha(a^* + ib^*)$ we have that the map α is surjective.

Let $x = a + ib$, $y = c + id$ for $a, b, c, d \in \Re$. Then

$$\alpha(xy) = \alpha((a+ib)(c+id)) = \alpha(ac - bd + i(bc + ad)) =$$

$$(ac - bd)^* + i(bc + ad)^* = (c^* + id^*)(a^* + ib^*) = \alpha(y)\alpha(x)$$

and

$$\alpha(x^*) = \alpha(a^* - ib^*) = a - ib = (a^* + ib^*)^* = \alpha(x)^*.$$

It is clear that the map α is a real linear. We shall show that α is complex linear. Let $x = a + ib$, $a, b \in \Re$. Then

$$\alpha(ix) = \alpha(i((a+ib)) = \alpha(-b + ia) =$$

$$-b + ia^* = i(a^* + ib^*) = i\alpha(x).$$

Since α is *-antiautomorphism of $\mathcal{U}(A)$, then it is an order isomorphism of $\mathcal{U}(A)$ onto itself.

Therefore the *-antiautomorphism α is a positive and normal. By Theorem 2 in [Dix 1, p. 56] it follows that the *-antiautomorphism α is a ultraweakly continuous. (for details see [S 4]).

Proposition 1.1.11. *The following conditions are equivalent:*

(1) *α is central, i.e. α acts identically on the center $Z_{\mathcal{U}}$ of the W^*-algebra $\mathcal{U}$;*

(2) $Z_{\mathcal{U}} = Z_A + iZ_A$;

(3) *A doesn't have direct summands isomorphic to the self-adjoint part of a W^*-algebra.*

Proof. (1) $\Rightarrow$ (2). Let α be central, i.e. α acts identically on the center of $\mathcal{U}$. If $x = a + ib \in Z_{\mathcal{U}}$, $a, b \in \Re(A)$, then from $\alpha(x) = x$ it follows that $a^* + ib^* = a + ib$, i.e. $a = a^*, b = b^*$, because $\Re(A) \bigcap \Re(A) = \{0\}$. Since $\Re(A)_s = A$, we have that $a, b \in A$ and a, b are central elements. Therefore $a, b \in Z_A$, i.e. $Z_{\mathcal{U}} = Z_A + iZ_A$.

(2) $\Rightarrow$ (1). Let $z \in Z_{\mathcal{U}}$. Then $z = a + ib$, with $a, b \in Z_A \subset \Re(A)$. Therefore by the definition of α we have $\alpha(z) = a^* + ib^* = a + ib = z$, i.e. α is central.

(2) $\Rightarrow$ (3). If $Z_{\mathcal{U}} = Z_A + iZ_A$, then $Z_A = (Z_{\mathcal{U}})_s$. Since the relation $Z_A \subset (Z_{\mathcal{U}})_s$ is always valid, the Corollary 1.1.10 implies that A doesn't have direct summands isomorphic to the self-adjoint part of a W^*-algebra.

(3) $\Rightarrow$ (2). Suppose that $e\mathcal{U} \bigcap A = \{0\}$ for some central projection e in $\mathcal{U}$. Then by Lemma 3.4 from [S 4] $(e\mathcal{U})_s = eA$, and there is a central projection g in A with $e \leq g$, such that the JW-algebras eA and gA are isomorphic, i.e. A contains the direct summand gA isomorphic to the self-adjoint part of the W^*-algebra $e\mathcal{U}$. This contradiction shows that $e\mathcal{U} \bigcap A \neq \{0\}$ for every central projection e in $\mathcal{U}$. Since $e\mathcal{U} \bigcap A$ is a weakly closed Jordan ideal in A, there is a central projection f in A such that $e\mathcal{U} \bigcap A = fA$, i.e. $0 \neq f \leq e$ (see [To 1, Proposition 5]). Thus given any projection $e \in Z_{\mathcal{U}}$ there exists a projection $f \in Z_A$ with $0 \neq f \leq e$. Since Z_A is weakly closed in $Z_{\mathcal{U}}$, this implies that the projection lattices in Z_A and $Z_{\mathcal{U}}$ coincide. Therefore $Z_{\mathcal{U}} = Z_A + iZ_A$. ∎

Let L be the lattices of all projections of JW-algebra A. L is said to be *modular* if $(e \vee f) \wedge g = e \vee (f \wedge g)$ for all $e, f, g \in L$ with $e \leq g$. A projection e in A is said to be *modular* if $[0, e] = \{f \in L : f \leq e\}$ is a modular lattices. A JW-algebra A is said to be

- *modular* if the lattices of all projections of A is modular, i.e. the projection $\mathbf{1}$ is modular;

- *locally modular* if A contains a faithful modular projection (i.e. A is a locally modular if its every direct summand contains a modular projection);

- *properly non-modular* if it has no nonzero central modular projection;

- *purely non-modular* if it has no nonzero modular projection;

Let $\Re$ be a real W^*-algebra, $\mathcal{U} = \Re + i\Re$ its enveloping W^*-algebra. If for two projections e, f in $A = \Re_s$ there exists an element x in $\Re$ (respectively x in $\mathcal{U}$) such that $e = x^*x$, $f = xx^*$, then we shall say that e and f are *equivalent* in $\mathfrak{R}$ (respectively in $\mathcal{U}$) and denote this by $e \sim_r f$ (respectively by $e \sim_c f$). If e and f are exchanged by a finite number of symmetries from A (i.e. $e = s_n ... s_1 f s_1 ... s_n$ for a finite family of symmetries $s_1, ..., s_n \in A$), then we call them *equivalent in* A and write $e \sim_s f$. It is clear that

$$(e \sim_s f) \Rightarrow (e \sim_r f) \Rightarrow (e \sim_c f). \tag{1.1}$$

Similarly to the complex case a projection e in a real W^*-algebra $\mathfrak{R}$ is said to be *finite* if for any projection $f \in \mathfrak{R}$, $f \leq e$, $f \sim_r e$ implies that $f = e$, i.e. e is not equivalent to its proper subprojection. A real W^*-algebra $\mathfrak{R}$ is *finite* if $\mathbf{1} \in \mathfrak{R}$ is a finite projection. Further in usual manner we define the notions of *semifinite*, *properly infinite*, *purely infinite* real W^*-algebras and aslo the types I_{fin}, I_∞, II_1, II_∞ and III for real W^*-algebras.

Proposition 1.1.12. *Let $\mathfrak{R}$ be a real or complex W^*-algebra. Then the JW-algebra $A = \mathfrak{R}_s$ is modular if and only if $\mathfrak{R}$ is finite, i.e. for $x \in \mathfrak{R}$, $x^*x = \mathbf{1}$ implies $xx^* = \mathbf{1}$.*

Proof. Let $A = \mathfrak{R}_s$ be modular, i.e.

$$(e \vee f) \wedge g = e \vee (f \wedge g),$$

for all projections e, f, g in A with $e \leq g$. And let $x \in \mathfrak{R}$, $x^*x = \mathbf{1}$. Then it is clear that $e = xx^*$ is also a projection in A. Suppose that $e \neq \mathbf{1}$, and put

$$p_1 = \mathbf{1} - e, \qquad p_n = x^n p_1 (x^*)^n, \quad n = 2, 3, ...$$

Since $p_n = u_n u_n^*$, $p_1 = u_n^* u_n$, where $u_n = x^n(\mathbf{1} - e) \in \mathfrak{R}, n = 2, 3, ...$, we obtain a sequence $\{p_n\}$ of orthogonal projections equivalent to p_1 in $\mathfrak{R}$. Note that if two orthogonal projections h, k in $A = \mathfrak{R}_s$ are equivalent in $\mathfrak{R}$, say $h = w^*w$, $k = ww^*$, $w \in \mathfrak{R}$, then they are exchanged by the symmetry $s = w + w^*(\mathbf{1} - h - k)$ in A. Therefore $\{p_n\}$ is a sequence of orthogonal projections in A any two of which are exchanged by a symmetry. This contradicts the modularity of A (see [To 1, Proposition 14]). Therefore $e = xx^* = \mathbf{1}$.

Conversely, let $\mathfrak{R}$ be finite, $e, f, g \in A = \mathfrak{R}_s$, $e \leq g$. Put

$$h = (e \vee f) \wedge g, \qquad k = e \vee (f \wedge g).$$

Then it is obvious that $h \geq k$ and moreover h and k are *perspective*, i.e., they have a common complement, because

$$h \vee f = k \vee f = e \vee f, \qquad h \wedge f = k \wedge f = g \wedge f.$$

By [H-OS, Proposition 5.2.3] (see also [To 1, Corollary 12]) there are two symmetries s_1, s_2 in A, such that $k = uhu^*$, where $u = s_1 s_2 \in \mathfrak{R}$. Put $x = hu^* + q \in \mathfrak{R}$, where $q = \mathbf{1} - h$. Then one has

$$x^* x = (uh + q)(hu^* + q) = k + q,$$

and on the other hand

$$xx^* = (hu^* + q)(uh + q) = h + hU^* q + quh + q = h + u^* kq + qku + q = \mathbf{1},$$

since $qk = kq = 0$. Therefore the finiteness of $\mathfrak{R}$ gives us that $k + q = \mathbf{1}$, i.e. $k = \mathbf{1} - q = h$. Thus $A = \mathfrak{R}_s$ is modular. ∎

Proposition 1.1.13. *Let $\mathfrak{R}$ be a real W^*-algebra such that $A = \mathfrak{R}_s$ is a modular JW-algebra. Then:*

(i) if $e, f \in A$, $e \leq f$, $e \sim_r f$, then $e = f$;

(ii) if $e \sim_r f$, then $\mathbf{1} - e \sim_r \mathbf{1} - f$.

Proof. Assertion (i) is obvious. Indeed, it is reduced to Proposition 1.1.12 by considering the real W^*-algebra $f\mathfrak{R}f$.

(ii) By the comparison theorem for JW-algebras [To 1, Theorem 10] (see also [H-OS, Theorem 5.2.13]) there exists a central projection z in A, such that

$$z(\mathbf{1} - e) \prec_s z(\mathbf{1} - f), \qquad (\mathbf{1} - z)(\mathbf{1} - e) \succ_s (\mathbf{1} - z)(\mathbf{1} - f),$$

Therefore there are projections h, k in A, such that

$$z(\mathbf{1} - e) \sim_s h \leq z(\mathbf{1} - f), \quad (\mathbf{1} - z)(\mathbf{1} - e) \sim_s k \leq (\mathbf{1} - z)(\mathbf{1} - f).$$

Thus (cf. [T 1, 13, Chap. V, Proposition 1.38]) we have from (1.1)

$$z = ze + z(\mathbf{1} - e) \sim_r zf + h \leq z,$$

$$\mathbf{1} - z = (\mathbf{1} - z)f + (\mathbf{1} - z)(\mathbf{1} - f) \sim_r (\mathbf{1} - z)e + k \leq \mathbf{1} - z.$$

Assertion (i) implies now that

$$h = z(\mathbf{1} - f), \qquad k = (\mathbf{1} - z)(\mathbf{1} - e).$$

Therefore

$$z(\mathbf{1} - e) \sim_r z(\mathbf{1} - f), \qquad (\mathbf{1} - z)(\mathbf{1} - e) \sim_r (\mathbf{1} - z)(\mathbf{1} - f),$$

i.e. $\mathbf{1} - e \sim_r \mathbf{1} - f$. ■

Theorem 1.1.14. *A reversible JW-algebra is modular if and only if its enveloping W^*-algebra is finite.*

Proof. Let A be a reversible JW-algebra, $\mathcal{U} = \mathcal{U}(A)$ its enveloping W^*-algebra.

If $\mathcal{U}$ is finite, then the projection lattice P of $\mathcal{U}$ is modular (see [T 1, Chap. V, Theorem 1.37]). By Theorem 1 in [To 1] the projection lattice of A is a sublattice of P and therefore it is also modular, i.e. the JW-algebra A is modular.

Conversely, let A be modular. We have to prove the finiteness of $\mathcal{U}$.

First, let $A \cong L^\infty(\Omega, \mu, V)$, where V is a finite-dimensional spin factor $(dim(V) \leq 6)$ (see Proposition1.1.2). Then $\mathcal{U}$ can be identified with the algebra $L^\infty(\Omega, \mu, \mathcal{U}(V))$, where $\mathcal{U}(V)$ is a finite-dimensional W^*-algebra of type I_n with $n < \infty$ (see [To 2, Theorem 2]). Therefore $\mathcal{U}$ is a finite W^*-algebra.

Now if the JW-algebra A *coincides* with the self-adjoint part of the W^*-algebra $\mathcal{U}$, then the assertion follows from Proposition 1.1.12. Therefore according Proposition 1.1.2 it is sufficient to consider the case when A is the self-adjoint part of the real W^*-algebra $\mathfrak{R} = \mathfrak{R}(A)$. In this case if A is isomorphic to the self-adjoint part of a W^*-algebra N, then by Proposition 1.1.8 $\mathcal{U}$ is *-isomorphic to the W^*-algebra $N + N_0$, where N_0

is the opposite W^*-algebra for N. Therefore Proposition 1.1.12 implies that N and hence $\mathcal{U}$ are finite W^*-algebras.

Thus from now on we may assume that A is not isomorphic to $L^\infty(\Omega, \mu, V)$ and has no direct summand isomorphic to the self-adjoint part of a W^*-algebra. Then according to Proposition 1.1.11 the involutive *-antiautomorphism $\alpha : \mathcal{U} \to \mathcal{U}$, defined as $\alpha(a + ib) = a^* + ib^*$, $a, b \in \Re$, acts identically on the center of $\mathcal{U}$. Note also that

$$A = \{x \in \mathcal{U}_s : \alpha(x) = x\}. \tag{1.2}$$

Assume that $\mathcal{U}$ is not finite. If a central projection z in A is finite with respect to $\mathcal{U}$, then $\mathcal{U}(zA) = z\mathcal{U}$ is a finite W^*-algebra. Therefore without loss of generality we suppose that A contains no nonzero central projections which are finite with respect to $\mathcal{U}$. Consider separately two alternative cases:

(a) $\mathcal{U}$ has a nonzero finite projection;

(b) $\mathcal{U}$ is of type III.

Case (a). Let e be a nonzero finite projection in $\mathcal{U}$. Since the involutive *-antiautomorphism α acts identically on the center of $\mathcal{U}$, by Lemma 3.3 from [S 3] the projection $\alpha(e)$ is equivalent to e and hence is also finite. Consider the projection $g = e \vee \alpha(e)$, which is finite by [T 1, Chap. V, Theorem 1.37]. One has that $g = s(e + \alpha(e))$, where $s(a)$ denotes the *range projection* of a self-adjoint operator a. According to (1.2) $e + \alpha(e) \in A$. Therefore $g \in A$ and g is finite with respect to $\mathcal{U}$. Consider a maximal family $\{g = g_1, g_2, ..., g_n\} \subset A$ of orthogonal projections any two of which are exchanged by a symmetry in A. Since A is modular, this family is finite [To 1, Theorem 11]. Put $f = \mathbf{1} - \sum_{i=1}^{n} g_i \in A$. By the comparison theorem for JW-algebras (see [To 1]) there exists a central projection z in A such that

$$zf \prec_s zg, \qquad (\mathbf{1} - z)f \;{}_s\!\succ (\mathbf{1} - z)g.$$

Moreover, $z \neq 0$ in view of the maximality of the family $\{g_i\}$. Therefore the projections zg_i, $n = 1, ..., n$ and $zf \prec_s zg$ all are finite with respect to $\mathcal{U}$, i.e., the nonzero central projection $z = \sum_{i=1}^{n} zg_i + zf$ from A is finite with respect to $\mathcal{U}$. This contradicts our assumption.

Case (b). Suppose that $\mathcal{U}$ is of type III. Let ρ be a normal state on $\mathcal{U}$, and let e denote the support projection of the normal state $\rho|_A$. Then $e\mathcal{U}e$ is a σ-finite (countably decomposable) W^*-algebra of type III. Since $e \in A$, passing, if necessary, to the algebras $e\mathfrak{R}e$, $eAe = (e\mathfrak{R}e)_s$, and $e\mathcal{U}e = e\mathfrak{R}e + ie\mathfrak{R}e$ instead of $\mathfrak{R}$, A, and $\mathcal{U}$, respectively, we may assume that $\mathcal{U}$ is a σ-finite W^*-algebra of type III.

Since A is not isomorphic to $L^\infty(\Omega, \mu, V)$, there exist three orthogonal projections e_1, e_2, e_3 in A pairwise equivalent in A. These projections have a common central cover z in A (see [To 1, Corollary 16]), which coincides with their central cover in $\mathcal{U}$ (see [S 2, Lemma 8.1]). Without loss generality we may suppose that $z = \mathbf{1}$. Since $\mathcal{U}$ is of type III and σ-finite [Sak 1, Proposition 2.2.14] implies that $e_1 \sim_c \mathbf{1}$, that is, $e_1 = u^*u$, $\mathbf{1} = uu^*$ for some u in $\mathcal{U} = \mathfrak{R} \oplus i\mathfrak{R}$. Let $u = x + iy$, $x, y \in \mathfrak{R}$. Then

$$\begin{aligned} e_1 &= x^*x \; + \; y^*y \\ \mathbf{1} &= xx^* \; + \; yy^* \end{aligned} \tag{1.3}$$

because $e_1, \mathbf{1} \in A = \mathfrak{R}_s$ and $\mathfrak{R} \bigcap i\mathfrak{R} = \{0\}$.

For $a \in B(H)$ denote by $r(a)$ and $l(a)$ respectivelythe *right* and *left projections* of a, which merge to the *range projection* $s(a)$ if $a^* = a$. In the general case $r(a) = s(a^*a)$, $l(a) = s(aa^*)$. Recall also that if $a = w|a|$ is the polar decomposition of a then $r(a) = w^*w$, $l(a) = ww^*$. If a lies in a real W^*-algebra $\mathfrak{R}$, then $w \in \mathfrak{R}$ by Lemma 1.1 from [S 5]. Therefore in this case $r(a), l(a) \in A = \mathfrak{R}_s$ and moreover $r(a) \sim_r l(a)$. Thus from (1.3) we obtain

$$\begin{aligned} e_1 &= s(x^*x) \vee s(y^*y) = r(x) \vee r(y) \\ \mathbf{1} &= s(xx^*) \vee s(yy^*) = l(x) \vee l(y) \end{aligned} \tag{1.4}$$

and also $r(x) \sim_r l(x)$, $r(y) \sim_r l(y)$.

By Proposition 1.1.13 (ii) one has $\mathbf{1} - r(x) \sim_r \mathbf{1} - l(x)$. Therefore (1.4) and [To 1, Corollary 8] imply

$$\mathbf{1} - e_1 \leq \mathbf{1} - r(x) \sim_r \mathbf{1} - l(x) = l(x) \vee l(y) - l(x)$$

$$\sim_s l(y) - l(x) \wedge l(y) \leq l(y) \sim_r r(y) \leq e_1 \sim_s e_2.$$

These combined with the formula (1.1) give us

$$\mathbf{1} - e_1 \prec_r e_2, \quad \text{i.e.} \;\; e_2 + e_3 \leq \mathbf{1} - e_1 \;\; \prec_r \;\; e_2.$$

This contradicts the assertion (i) of Proposition 1.1.13.

Thus in the both Cases (a) and (b) the assumption of the infiniteness of $\mathcal{U}$ contradicts the modularity of A. ∎

It should be noted that in this proof the condition $\mathfrak{R} = \mathfrak{R}(A)$ was not essential, but only the property $A = \mathfrak{R}_s$ was used. Therefore Theorem 1.1.14 can be reformulated as follows (cf. [St 4, Lemma 2.4]).

Theorem 1.1.15. *Let* $\mathfrak{R}$ *be a real* W^**-algebra,* $\mathcal{U} = \mathfrak{R} + i\mathfrak{R}$ *its enveloping* W^**-algebra. Then* $\mathfrak{R}$ *is finite if and only if* $\mathcal{U}$ *is finite.*

For locally modular JW-algebras we have the following result (cf. [A 1, Theorem 4]).

Corollary 1.1.16. *If a reversible* JW*-algebra* A *is locally modular then its enveloping* W^**-algebra* $\mathcal{U} = \mathcal{U}(A)$ *is semifinite.*

Proof. As in Theorem 1.1.14, we may assume that $\mathfrak{R} \bigcap i\mathfrak{R} = \{0\}$. If A is locally modular, then it contains a modular projection e with central cover $\mathbf{1}$. As it was mentioned above the central cover of e with respect to $\mathcal{U}$ is also $\mathbf{1}$. Consider the modular JW-algebra $eAe = (e\mathfrak{R}e)_s$, where $\mathfrak{R} = \mathfrak{R}(A)$. By Proposition 1.1.12 $e\mathfrak{R}e$ is a finite real W^*-algebra and thus by Theorem 1.1.14 W^*-algebra $e\mathcal{U}e$ is also finite. Therefore e is a finite projection in $\mathcal{U}$ with the central cover $\mathbf{1}$, i.e. $\mathcal{U}$ is semifinite. ∎

1.2. Traces on JW-algebras

Definition 1.2.1. A *trace* on a JW-algebra A (respectively on a real W^*-algebra $\mathfrak{R}$) is a function τ on the set A^+ of positive elements of A (respectively on the set $\mathfrak{R}^+$ of positive elements of $\mathfrak{R}$) with values in $[0, +\infty]$, satisfying the following conditions:

1) $\tau(x + y) = \tau(x) + \tau(y)$, for all $x, y \in A^+$ (respectively $x, y \in \mathfrak{R}^+$)
2) $\tau(\lambda x) = \lambda\tau(x)$, for all $x \in A^+$ (respectively $x \in \mathfrak{R}^+$) and $\lambda \in \mathbf{R}$, $\lambda \geq 0$, where $0 \cdot (+\infty) = 0$;
3) $\tau(sxs) = \tau(x)$, for an arbitrary symmetry s and x in A (respectively for each symmetry s and x in $\mathfrak{R}$).

The trace τ is said to be *faithful*, if $\tau(x) > 0$ for all nonzero $x \in A^+$; *finite* if $\tau(\mathbf{I}) < +\infty$; *semifinite* if given any $x \in A^+$ there is a nonzero $y \in A^+$, $y \leq x$ with $\tau(y) < +\infty$. The trace τ is *normal* if for every net $\{x_\alpha\}$ monotone increasing and bounded in A^+ we have $\tau(\sup x_\alpha) = \sup \tau(x_\alpha)$.

Proposition 1.2.2. *Let $\mathfrak{R}$ be a real W^*-algebra and τ a trace on the JW-algebra $A = \mathfrak{R}_s$. Then $\tau(x^*x) = \tau(xx^*)$ for all $x \in \mathfrak{R}$.*

Proof. Let $u \in \mathfrak{R}$ be an arbitrary unitary operator. Consider the $*$-automorphism α of the JW-algebra A defined as $\alpha(a) = u^*au$, $a \in A$. Since $A = \mathfrak{R}_s$, it is clear that A is a reversible JW-algebra. The proof of Theorem 1.1.4 from [Up] shows that the $*$-automorphism α is a finite product of involutions $U_s : a \to sas$, where s is a symmetry in A. Thus by the property 3) of traces $\tau(u^*au) = \tau(a)$ for all $a \in A^+$, u - unitaries from $\mathfrak{R}$.

Now let x be an arbitrary element from $\mathfrak{R}$. Without loss of generality we may assume that $\|x\| < 1/2$. Then the operator $\mathbf{I}+x$ is invertible and hence in its polar decomposition $(\mathbf{I}+x) = u|\mathbf{I}+x|$, u is a unitary operator. By Lemma 1.1 from [S 5] this unitary u lies in the real W^*-algebra $\mathfrak{R}$. Therefore by the previous paragraph we have:

$$\tau((\mathbf{I} + x)^*(\mathbf{I} + x)) = \tau(|\mathbf{I} + x|uu^*|\mathbf{I} + x|) = \tau(|\mathbf{I} + x|^2)$$

$$= \tau(u|\mathbf{I} + x|^2 u^*) = \tau((\mathbf{I} + x)(\mathbf{I} + x)^*),$$

i.e.

$$\tau(x^*x + x + x^* + \mathbf{I}) = \tau(xx^* + x + x^* + \mathbf{I}).$$

Since $\|x + x^*\| \leq \|x\| + \|x^*\| < 1$ and the element $x + x^*$ is contained in $\mathfrak{R}_s = A$ the element $x^* + x + \mathbf{I}$ is a positive element from A. Hence by the property 1) of traces:

$$\tau(x^*x + x + x^* + \mathbf{I}) = \tau(x^*x) + \tau(x + x^* + \mathbf{I})$$
$$= \tau(xx^* + x + x^* + \mathbf{I}) = \tau(xx^*) + \tau(x + x^* + \mathbf{I}).$$

Therefore $\tau(x^*x) = \tau(xx^*)$ for all $x \in \mathfrak{R}$. ■

Theorem 1.2.3. *A JW-algebra A is modular if and only if it possesses a separating family of finite normal traces.*

Proof. Let $\{\tau_\nu\}$ be a separating family of finite normal traces on A, and let e, f, g be projections in A with $e \leq g$. Put

$$h = (e \vee f) \wedge g, \qquad k = e \vee (f \wedge g).$$

Then $k \leq h$ and $k = s_1 s_2 h s_2 s_1$ with s_1 and s_2 symmetries from A (see the proof of Proposition1.1.12). Therefore $h - k \in A^+$ and $\tau_\beta(h - k) = 0$ for all β. Thus $h - k = 0$, and hence A is modular.

Conversely, let A be modular JW-algebra. If A is reversible, then by Theorem 1.1.14 its enveloping W^*-algebra $\mathcal{U}$ is finite and hence possesses a separating family $\{\tau_\nu\}$ of finite normal traces [T 1, Chap. V, Theorem 2.4]. Then $\{\tau_\nu|_A\}$ gives a separating family of finite normal traces on A. If A not reversible, then by Proposition 1.1.2 we may assume that $A = L^\infty(\Omega, \mu, V)$, where μ is a Radon measure on a locally compact space Ω and V is a spin factor. Let t be the canonical normalized trace on V [To 2]. For $f \in A$, $f : \Omega \to V$, put

$$\tau(f) = \int_\Omega t(f(\omega))\mu(d\omega).$$

It is easy to see that τ is a faithful finite normal trace on A. ■

This theorem combined with Propositions 1.1.12 and 1.2.2 implies

Corollary 1.2.4. *A real W^*-algebra $\mathfrak{R}$ is finite if and only if it possesses a separating family of finite normal traces.*

It should be noted that Corollary 1.2.4 follows also from Theorem 1.1.15. Another proof of this result for real factors was suggested in [St 4, Lemma 2.3].

Proposition 1.2.5. *Let A be a reversible JW-algebra, τ_1 a semifinite trace on the enveloping W^*-algebra $\mathcal{U}(A)$. Then its restriction $\tau = \tau_1|_A$ is a semifinite trace on A.*

Proof. As we noted in above one can consider only a purely real JW-algebra A. In its turn this case is reduced into two alternative cases:

(a) A is isomorphic to the self-adjoint part of a W^*-algebra;

(b) A doesn't have direct summands isomorphic to self-adjoint part of a W^*-algebra.

Consider the case (a) and let A be isomorphic to N_s where N is a W^*-algebra. By Proposition 1.1.8 and Remark 1.1.7 $\mathcal{U}$ can be identified with the W^*-algebra $N \oplus N_0 = \{(a, b_0),\ a, b \in N\}$ and $A = \{(a, a_0),\ a \in N_s\}$. Let (a, a_0) be an arbitrary positive non zero element from A. We must find a non zero element (c, c_0) in A with $0 < (c, c_0) \leq (a, a_0)$ and $\tau((c, c_0)) < +\infty$. Consider the element $(a, 0) \in N \oplus N_0 = \mathcal{U}$. The semifiniteness of τ_1 on $\mathcal{U}$ implies the existence of a non zero $b \in N$ such that $0 \leq (b, 0) \leq (a, 0)$ with $\tau_1((b, 0)) < +\infty$. Similarly for the element $(0, b_0) \in \mathcal{U}$ there exists a non zero $c \in N$ such that $0 \leq (0, c_0) \leq (0, b_0)$ and $\tau_1((0, c_0)) < +\infty$. It is clear now that

$$(c, c_0) \neq 0, \quad 0 \leq (c, c_0) \leq (b, b_0) \leq (a, a_0)$$

and

$$\tau((c, c_0)) = \tau_1((c, 0)) + \tau_1((0, c_0)) \leq \tau_1((b, 0)) + \tau_1((0, c_0)) < +\infty.$$

Therefore τ is semifinite.

Consider now the case (b). In this case the involutive*-antiautomorphism α of $\mathcal{U}$, generating A, is central by Proposition 1.1.11. Therefore

[S 3, Lemma 3.3] implies that given any projection $e \in \mathcal{U}$ the projections e and $\alpha(e)$ are equivalent. Thus

$$\tau_1(e) = \tau_1(\alpha(e)) = \tau(\frac{1}{2}(e + \alpha(e))),$$

where $\frac{1}{2}(e + \alpha(e))$ is a non zero positive element in A.

Now let e_0 be any projection in A. By the assumption there is a non zero projection $e \in \mathcal{U}$ with $0 \leq e \leq e_0$, $\tau_1(e) < +\infty$. Put $p = \frac{1}{2}(e + \alpha(e))$. Then $\tau(p) = \tau_1(e) < +\infty$ and $p \leq \frac{1}{2}(e_0 + \alpha(e_0)) = e_0$ since $\alpha(e_0) = e_0 \in A$. Therefore given any non zero projection $e_0 \in A$ we can find a non zero element $p \in A$ with $0 \leq p \leq e_0$ and $\tau(p) < +\infty$, i.e. τ is semifinite. ■

Now we can give an alternative proof of the following result of [A 1, Thorem 5].

Theorem 1.2.6. *A JW-algebra A is locally modular if and only if it possesses a faithful semifinite normal trace.*

Proof. The "if" part is clear. The proof of the "only if" part for the case when A is isomorphic to the direct sum of JW-algebras of the form $L^\infty(\Omega, \mu, V)$ is similar to Theorem 1.2.3 Therefore it is suffcient to show the existence of a faithful semifinite normal trace on a reversible locally modular JW-algebra A. But in this case the enveloping W^*-algebra $\mathcal{U}$ is semifinite (Corollary 1.1.16) and hence possesses a faithful semifinite normal trace τ. According to Proposition1.2.5 its restriction $\tau|_A$ is a faithful semifinite normal trace on A. ■

Corollary 1.2.7. *A real W^*-algebra is semifinite if and only if it has a faithful semifinite normal trace.*

Theorem 1.2.8. *A JW-algebra A is purely non modular if and only if there is no non-null semifinite normal trace on A.*

Proof. Let A be a purely non modular JW-algebra and τ a semifinite normal trace on A with the support $s(\tau)$. Then the JW-algebra $s(\tau) \cdot A$

possesses a faithful semifinite normal trace and hence it is locally modular by Theorem 1.2.6. Thus $s(\tau) = 0$, i.e. $\tau = 0$.

Conversely, let A be a JW-algebra without any non-null semifinite normal trace and $A = eA + (\mathbf{1} - e)A$ its decomposition into the locally modular and purely non modular portions. Then eA admits a faithful semifinite normal trace τ by Theorem 1.2.6. Then $\tau_0(x) = \tau(ex)$, $x \in A^+$, is a semifinite normal trace on A with $s(\tau_0) = e$. Thus $e = 0$ and A is a purely non modular JW-algebra. ∎

Theorem 1.2.9. *Let $\mathfrak{R}$ be a real W^*-algebra, $\mathcal{U} = \mathfrak{R} + i\mathfrak{R}$ its enveloping W^*-algebra. And let τ be a normal trace on the JW-algebra $A = \mathfrak{R}_s$. Then τ can be extended to a normal trace τ_1 on the W^*-algebra $\mathcal{U}$. If τ is faithful, then τ_1 is also faithful. If τ is finite (respectively semifinite), then τ_1 is finite (respectively semifinite).*

Proof. Let $x = a + ib \in \mathcal{U}^+$, where $a, b \in \mathfrak{R}$. It is easy to see that $a \in \mathfrak{R}^+ = A^+$. Put $\tau_1(x) = \tau(a)$. Since $\mathfrak{R} \bigcap i\mathfrak{R} = \{0\}$, the function $\tau_1 : \mathcal{U}^+ \to [0, +\infty]$, is well defined and satisfies the conditions:

1) $\tau_1(x + y) = \tau_1(x) + \tau_1(y)$ for $x, y \in \mathcal{U}^+$;

2) $\tau_1(\lambda x) = \lambda \tau_1(x)$ for $x \in \mathcal{U}^+$, $\lambda \in \mathbf{R}$, $\lambda \geq 0$.

In order to prove that τ_1 is a trace we must show that $\tau_1(x^*x) = \tau_1(xx^*)$ for all $x \in \mathcal{U}$. Let $x = a + ib$, $a, b \in \mathfrak{R}$. Then

$$x^*x = (a^* - ib^*)(a + ib) = a^*a + b^*b + i(a^*b - b^*a).$$

Thus by the definition of τ_1: $\tau_1(x^*x) = \tau(a^*a) + \tau(b^*b)$. Similarly $\tau_1(xx^*) = \tau(aa^*) + \tau(bb^*)$. Since a and b lie in $\mathfrak{R}$, $\tau(a^*a) = \tau(aa^*)$, $\tau(b^*b) = \tau(bb^*)$ by Proposition 1.2.2. Therefore $\tau_1(x^*x) = \tau_1(xx^*)$, i.e. τ_1 is a trace on the W^*-algebra $\mathcal{U}$.

Suppose that τ is faithful and $\tau_1(x) = 0$, where $x = a + ib \in \mathcal{U}^+$. Then by Corollary 1.1.4 $a \in A^+$ and $\tau(a) = \tau_1(x) = 0$. Since τ is faithful, $a = 0$ and hence $x = ib \in \mathcal{U}^+$. Then by Lemma 1.1.3 $ib = 0$, i.e. $x = a + ib = 0$. Therefore τ_1 is also faithful.

If τ is finite then obviously τ_1 is also finite.
The normality of τ_1 is checked directly.

Finally let τ be semifinite. By [T 1, Chap. V, Lemma 2.13] there is a central projection $z \in \mathcal{U}$ such that τ_1 is semifinite on $z\mathcal{U}$ and $\tau_1(x) = +\infty$ for all non zero $x \in (\mathbf{1} - z)\mathcal{U}^+$. Suppose that $\mathbf{1} - z > 0$. Then $\mathbf{1} - z = a_o + ib_o$ with $a_o, b_o \in \Re$, and $a_o \in A^+$ since $\mathbf{1} - z > 0$. Since z is central and $\Re \bigcap i\Re = \{0\}$, a_o is a central element of A. Given any non zero element $y \in A^+$, $y \leq a_o$ we have

$$(\mathbf{1} - z)y = a_o y + ib_o \neq 0$$

and hence

$$\tau(a_o y) = \tau_1((\mathbf{1} - z)y) = +\infty.$$

On the other hand

$$\tau(a_o y) = \tau\left(y^{1/2} a_o y^{1/2}\right) \leq \|a_o\| \tau(y).$$

Therefore $\tau(y) = +\infty$ for all non zero y in A^+, with $y \leq a_o$. This contradicts the semifiniteness of τ. Thus $\mathbf{1} - z = 0$, i.e. τ_1 is semifinite. ∎

Corollary 1.2.10. *Let A be a reversible JW-algebra with a normal trace τ, $\mathcal{U} = \mathcal{U}(A)$ its enveloping W^*-algebra. Then the trace τ can we extended to a normal trace τ_1 on $\mathcal{U}$. If τ is faithful (respectively finite or semifinite) then τ_1 is also faithful (respectively finite or semifinite).*

Remark 1.2.11. The reversibility condition in Corollary 1.2.10 can not be omitted. Indeed, let A be an infinite dimensional spin factor and let τ_0 be the canonical trace on A. Then the C^*-algebra B, generated by A is prime and separable and it has a unique normal trace τ which extends the canonical trace τ_o. For every state ρ on B let π_ρ be the GNS-representation of B in a Hilbert space H_ρ with a cyclic vector $\xi_\rho \in H_\rho$. Since B is a simple C^*-algebra, then π_ρ is an isomorphism and $\pi_\rho(A)$ generates $\pi_\rho(A)''$ as W^*-algebra. It is known that the C^*-algebra B of the canonical anti-commutative relations have represent of types I_∞, II_1, II_∞ and III [E]. Thus dependening on the state ρ actions the spin factor $\pi_\rho(A)$ which isomorphic to A, can generate W^*-algebra of all types. In particular, a trace τ_o on $\pi_\rho(A)$ can not be extended to a trace on the enveloping W^*-algebra. But if the state ρ on B, is a trace $\rho = \tau$, then the W^*-algebra $\pi_\tau(A)''$ which constructed by the GNS-representation π_ρ is an injective W^*-factor of type II_1 and the unique trace on $\pi_\tau(A)''$ is the extention of the trace τ_o on $\pi_\tau(A)$.

Theorem 1.2.9 and Corollary 1.2.10 mean that to any trace on a reversible JW-algebra A corresponds a trace on the enveloping W^*-algebra $\mathcal{U}$ (its extension) and conversely, to any trace on $\mathcal{U}$ corresponds a trace on A (its restriction). Moreover this correspondence preserves such properties as normality, faithfulness, finiteness and semifiniteness. Now we shall consider the question: when this correspondence is *one-to-one*, i.e. when the extension constructed in Theorem 1.2.9 is unique. The answer is given by the following theorem.

Theorem 1.2.12. *Let A be a purely real JW-algebra, $\mathcal{U}$ its enveloping W^*-algebra. Each of the conditions (1) - (3) in Proposition 1.1.11 is equivalent to the following*

(4) any trace on A has a unique extension to a trace on $\mathcal{U}$.

If moreover A is locally modular, then we have one more equivalent condition:

(5) any normal semifinite trace on A has a unique extension to a normal semifinite trace on $\mathcal{U}$.

Proof. It is sufficient to show the implications (1) $\Rightarrow$ (4) $\Rightarrow$ (5), (4) $\Rightarrow$ (2) and if A is locally modular then (5) $\Rightarrow$ (2).

We shall show that (1) $\Rightarrow$ (4). If the *-antiautomorphism α, generating A, is central then from the proof of Proposition 1.2.5 we know that given any trace τ_1 on $\mathcal{U}$ the equality

$$\tau_1(e) = \tau(\frac{1}{2}(e + \alpha(e))),$$

holds for all projections e from $\mathcal{U}$, where τ is the restriction of τ_1 to A. This means that the trace τ_1 on $\mathcal{U}$ is completely defined by its values on A. Namely $\tau_1 = \tau_0$, where τ_0 is the extension of the trace τ on A constructed in Theorem 1.2.9. Therefore any trace on A has a unique extension.

(4) $\Rightarrow$ (5). If τ is an arbitrary normal semifinite trace on A, then by assumption it has a unique extension which necessarily coincides with τ_0 and therefore is normal and semifinite by Theorem 1.2.9.

(4) $\Rightarrow$ (2). Suppose that (2) is not valid, i.e. $Z_{\mathcal{U}} \neq Z_A + iZ_A$. Then there is a non zero projection $p \in Z_{\mathcal{U}}$ ($\neq 0$), which doesn't majorize any non zero projection from Z_A. We claim that $p\alpha(p) = 0$. In fact, $p\alpha(p)$ is a central projection in $\mathcal{U}$ and $\alpha(p\alpha(p)) = p\alpha(p)$, i.e. $p\alpha(p) \in Z_A$. Since $p\alpha(p) \leq p$ it follows that $p\alpha(p) = 0$. Let ρ be any state on $Z_{\mathcal{U}}$ such that $\rho(p) \neq 0$. By the Approximation Theorem [Dix 1, Theorem 1, p. 272] any semifinite W^*-algebra possesses a center-valued trace, say Φ (see also [To 1, Corollary 28]). Put $\tau_1 = \rho \circ \Phi$. Then τ_1 is a trace on $\mathcal{U}$ and $\tau_1(p) = \rho(\Phi(p)) = \rho(p) \neq 0$. Put now

$$\tau_2(x) = \tau_1(p^{\perp}x + p\alpha(x)), \qquad x \in \mathcal{U}^+.$$

Since $p \in Z_{\mathcal{U}}$ and α is a *-antiautomorphism, the map $\tau_2 : \mathcal{U}^+ \to [0, +\infty]$ is a trace. If $x \in A^+$, then $\alpha(x) = x$ and hence $\tau_2(x) = \tau_1(x)$, i.e. $\tau_2|_A = \tau_1|_A$. At the same time $\tau_2(p) = \tau_1(p^{\perp}p + p\alpha(p)) = 0$ and $\tau_1(p) \neq 0$. Therefore $\tau_1 \neq \tau_2$. This means that the trace $\tau = \tau_1|_A = \tau_2|_A$ has two different extension to $\mathcal{U}$ contradicting to (4).

Finally, let A be locally modular. Then by Corollary 1.1.16 the W^*-algebra $\mathcal{U}$ semifinite and hence possesses a separating set of normal semifinite traces.

If (2) doesn't hold, then as we know from above there is a non zero projection $p \in Z_{\mathcal{U}}$ such that $p\alpha(p) = 0$. Let τ_1 be a normal semifinite trace on $\mathcal{U}$ with $\tau_1(p) \neq 0$. Putting

$$\tau_2(x) = \tau_1(p^{\perp}x + p\alpha(x)), \qquad x \in \mathcal{U}^+,$$

we obtain a normal semifinite trace τ_2 such that $\tau_2|_A = \tau_1|_A$ and $\tau_2(p) = 0$, while $\tau_1(p) \neq 0$. This contradicts to the condition (5). Therefore, if A is locally modular, then (5) $\Rightarrow$ (2). ∎

If the JW-algebra A is modular, then by Theorem 1.1.14 $\mathcal{U}(A)$ is a finite W^*-algebra and hence possesses a separating set of normal finite traces. Therefore similar to the locally modular case one can easily prove the following result.

Corollary 1.2.13. *Let A be a modular purely real JW-algebra. Then the conditions (1) - (3) from Proposition 1.1.11 and the conditions (4), (5) from Theorem 1.2.12 are equivalent to the following*

(6) any normal finite trace on A has a unique extension to a normal finite trace on the enveloping W^*-algebra $\mathcal{U}$.

1.3. Types of JW-algebras and enveloping von Neumann algebras

Recall the following classification of JW-algebras suggested by D.Topping [To 1]:

we say that a JW-algebra A has:

- the *type* I if it contains a faithful abelian projection, i.e. if there is a faithful projection e in A such that eAe is an abelian JW-algebra;

- the *type* I_{fin}, if it is of type I and modular;

- the *type* I_∞, if it is of type I, properly non modular;

- the *type* II, if it is locally modular and doesn't contain nonzero abelian projections;

- the *type* II_1, if it is of type II and modular;

- the *type* II_∞, if it is of type II, properly non modular;

the *type* III, if it is purely non modular.
(see [A 4],[A 9],[S 2],[To 1]).

In [To 1, Theorem 13] Topping has proved that given any JW-algebra A there exist five uniquely defined orthogonal central projections $e_1, e_2, \dots, e_5 \in A$ such that $e_1 + e_2 + \dots + e_5 = \mathbf{1}$ and

$$e_1 A \text{ is a } JW\text{-algebra of type } \mathrm{I}_{fin},$$

$$e_2 A \text{ is a } JW\text{-algebra of type } \mathrm{I}_\infty,$$

$$e_3 A \text{ is a } JW\text{-algebra of type } \mathrm{II}_1,$$

$$e_4 A \text{ is a } JW\text{-algebra of type } \mathrm{II}_\infty,$$

$$e_5 A \text{ is a } JW\text{-algebra of type III},$$

and a JW-factor has one and only one of these types.

Having the classifications of JW-algebras and W^*-algebras it is natural to study the relations between the types of JW-algebras and their enveloping W^*-algebras. In the particular case, when our JW-algebra happens to be the self-adjoint part of a W^*-algebra the type decomposition coincides with the classical one. This was mentioned by Topping in [To 1, p. 30]. But this question is not obvious if we pass to general JW-algebras. The first result in this direction was obtained by Stormer. He proved the following result (see [S 2, Corollary 6.5].

Theorem 1.3.1. *Let A be a reversible JW-algebra and $\mathcal{U}$ its enveloping W^*-algebra. Then A is of type I if and only if $\mathcal{U}$ is of type I.*

Now using Theorems 1.2.8 and 1.3.1, we shall prove the main result of this Section.

Theorem 1.3.2. *A reversible JW-algebra A is of type I_{fin}, I_∞, II_1, II_∞ or III if and only if its an enveloping W^*-algebra $\mathcal{U}$ has the corresponding type.*

Proof. If A is a reversible JW-algebra of type I_{fin}, then by Theorem 1.1.14 it follows that $\mathcal{U}$ is a finite W^*-algebra and by Theorem 1.3.1 $\mathcal{U}$ is of type I_{fin}.

Conversely, let $\mathcal{U}$ be a W^*-algebra of type I_{fin}. Theorem 1.1.14 implies that A is a modular JW-algebra. By Theorem 1.3.1 A is of type I_{fin}.

Let A be of type I_∞. Since A is reversible, by Theorem 1.3.1 $\mathcal{U}$ is of type I. We shall show that $\mathcal{U}$ is properly infinite. Let e be a central finite projection in $\mathcal{U}$. It is clear that $e\mathcal{U} \bigcap A$ is a modular JW-algebra. Since e is a central projection, then $e\mathcal{U}$ is a weakly closed two-sided ideal in $\mathcal{U}$ and hence $e\mathcal{U} \bigcap A$ is a weakly closed Jordan ideal in A. By [To 1, Corollary 2 and Proposition 5] there exists a central projection e_1 in A such that

$$e\mathcal{U} \bigcap A = e_1 A.$$

Since $e_1 \leq e$, then $e_1\mathcal{U} = e_1(e\mathcal{U})$ is a finite W^*-algebra. But $e_1\mathcal{U} = \mathcal{U}(e_1 A)$ is the W^*-algebra generated by $e_1 A$. At the same time $e_1 A$ is reversible JW-algebra which is properly nonmodular. This contradicts Theorem 1.1.14. Therefore $e_1 = 0$, i.e. $e\mathcal{U} \bigcap A = \{0\}$.
Thus we can apply [S 4, Lemma 3.4] (see also [A 9, Proposition 3.6]) to the central projection e. By that Lemma there exist a central projection $g \in A$, $g \geq e$ and the Jordan isomorphism $\psi : gA \to (e\mathcal{U})_s$. Since g is a non modular projection, then there exists an infinite sequence $\{g_i\}$ of orthogonal projections in gA any two of which are exchanged by a symmetry in gA (see [To 1, Lemma 2.3]).

Then $\{\psi(g_i)\}$ is an infinite sequence of orthogonal projections in $e\mathcal{U}$, any two of which are exchanged by a symmetry (therefore are equivalent) in $e\mathcal{U}$. This contradicts the finiteness of e. Hence $e = 0$, i.e. $\mathcal{U}$ is properly infinite and therefore it is of type I_∞.

Conversely, let $\mathcal{U}$ is of type I_∞. By Theorem 1.1.14 if e is a central modular projection in A, then e is a central finite projection in $\mathcal{U}$ and therefore $e = 0$. Hence A is a properly non modular JW-algebra. By Theorem 1.3.1 A is of type I, i.e. JW-algebra A is of type I_∞.

Let A be a JW-algebra of type II_1. By Theorem 1.1.14 $\mathcal{U}$ is a finite W^*-algebra. We shall show that $\mathcal{U}$ does not contain any non zero abelian projections. If $A = \mathcal{U}(A)_s$, then the assertion is trivial. Therefore let A be a purely real JW-algebra. Let $\mathcal{U} = e\mathcal{U} + (\mathbf{1} - e)\mathcal{U}$ be the decomposition of $\mathcal{U}$ into the direct sum of the subalgebras $e\mathcal{U}$ of type I and $(\mathbf{1} - e)\mathcal{U}$ without nonzero abelian projections, where $e \in Z_{\mathcal{U}}$.

Since e is central $e\mathcal{U}$ is a weakly closed two sided ideal in $\mathcal{U}$ and thus $e\mathcal{U} \bigcap A$ is a weakly closed Jordan ideal in A. Therefore $e\mathcal{U} \bigcap A$ for an appropriate central projection $e_1 \in A$. Since $e_1 \leq e$, one has that $e\mathcal{U} = e_1(e\mathcal{U})$ is a W^*-algebra of type I, which is generated by the JW-algebra $e_1 A$ of type II. This contradicts Theorem 1.3.1 and therefore $e_1 = 0$, i.e. $e\mathcal{U} \bigcap A = \{0\}$. Again by [S 4, Lemma 3.4] there exist a central projection $g \in A$, $g \geq e$, and a Jordan isomorphism ψ between $(e\mathcal{U})_s$ and gA. Since $e\mathcal{U}$ is of type I, there exists an abelian projection $p \in \mathcal{U}$, $p \neq 0$, i.e. $p(e\mathcal{U})p$ is an abelian W^*-algebra. Therefore

$$g = \psi(e) \geq \psi(p) \neq 0$$

and $\psi(p)A\psi(p)$ is an abelian JW-algebra, i.e. $\psi(p)$ is an abelian projection in A. Contradiction shows that $e = 0$, i.e. $\mathcal{U}$ is of type II_1.

Conversely, let $\mathcal{U}$ be a W^*-algebra of type II_1. By Theorem 1.1.14 A is a modular JW-algebra. By the type decomposition theorem for JW-algebra there is a central projection $e \in A$ such that eA is of type I_{fin}, and $(\mathbf{1} - e)A$ is of type II_1. By the assumption the JW-algebra A is reversible. Therefore eA is also reversible JW-algebra and of type I. By Theorem 1.3.1 its enveloping W^*-algebra $e\mathcal{U} = \mathcal{U}(eA)$ is of type I. On the other hand $\mathcal{U}$ and hence $e\mathcal{U}$ are of type II_1. This contradiction shows that $e = 0$, i.e. A is of type II_1.

Before going on with the proof of Theorem 1.3.2 we give one result of preliminary nature.

Let $\mathcal{U}$ be an arbitrary W^*-algebra, K its state space, i.e. the set of all normal states on $\mathcal{U}$. Recall that on (norm) bounded subsets of $\mathcal{U}$ the weak and the ultraweak topologies coincide and are given by the family of seminorms $\{|\omega(x)|, \omega \in K\}$, where $x \in \mathcal{U}$. On bounded sets the strong and the ultrastrong topologies also coincide and are determined by the family of seminorms $\{[\omega(x^2)]^{1/2}, \omega \in K\}$, where $x \in \mathcal{U}$. For details see [T 2, Chap. V, Lemma 2.5].

Below speaking about these topologies on a JW-algebra A or on its enveloping real algebra $\Re(A)$ we shall mean that they are restricted from the enveloping W^*-algebra $\mathcal{U}$.

Lemma 1.3.3. *Let A be a JW-algebra which is not modular, $\mathfrak{R} = \mathfrak{R}(A)$ its enveloping weakly closed real algebra. Then the map $x \to x^*$ is strongly discontinuous on the unit ball of $\mathfrak{R}$.*

Proof. Since $\mathbf{1}$ is not a modular projection there exists an infinite orthogonal sequence $\{g_i\}$ of projections any two of which are exchanged by a symmetry in A. Let $g_1 = s_i g_i s_i$, $i = 1, 2, \ldots$, where s_i is a symmetry in A. Consider the sequence $\{t_i = s_i g_i\}$ in $\mathfrak{R}$. We have

$$t_i t_i^* = s_i g_i s_i = g_i, \quad t_i^* t_i = g_i s_i^2 g_i = g_i, \qquad i = 1, 2, \ldots.$$

If ω is an arbitrary normal state from K then

$$\sum_{i=1}^{\infty} \omega(g_i) = \omega\left(\sum_{i=1}^{\infty} g_i\right) \leq \omega(\mathbf{1}) = 1.$$

Thus $\omega(t_i^* t_i) = \omega(g_i) \to 0$, as $t \to \infty$, i.e. $t \to 0$ strongly. At the same time $\omega(t_i t_i^*) = \omega(g_1)$ for all $i = 1, 2, \ldots$, and hence $\{t_i^*\}$ doesn't converge to zero strongly. Noticing that $\|t_i\| = \|s_i g_i\| \leq 1$, for all $i = 1, 2, \ldots$ we finish the proof. ∎

Now we shall go on with the proof Theorem 1.3.2, and firstly prove for the case of JW-algebra of type III.

Let A be a JW-algebra of type III. Since A is reversible we may assume that A is a purely real JW-algebra i.e. $\mathfrak{R}(A) \bigcap i\mathfrak{R}(A) = \{0\}$, and $\mathcal{U} = \mathcal{U}(A) = \mathfrak{R}(A) + i\mathfrak{R}(A)$. Consider the map $\Phi : \mathcal{U} \to \mathfrak{R}(A)$ defined as $\Phi(x + iy) = x$, where $x, y \in \mathfrak{R}(A)$. By Corollary 1.1.4 Φ is a positive linear map. It is easy to show that Φ is a ultraweakly continuous and hence weakly continuous on bounded subsets of $\mathcal{U}$. In virtue of $\mathfrak{R}(A) \bigcap i\mathfrak{R}(A) = \{0\}$ it is clear that $\Phi(xuy) = x\Phi(u)y$ for all $x, y \in \mathfrak{R}(A)$, $u \in \mathcal{U}$.

Suppose that $\mathcal{U}$ is not of type III. Then there is a non zero central projection $z \in \mathcal{U}$ such that $z\mathcal{U}$ is a semifinite W^*-algebra and hence possesses a faithful semifinite normal trace τ (see [T 1, Chap. V, Theorem 2.15]).

Let $a \in z\mathcal{U}$ be a positive element with $\tau(a^2) < +\infty$. Since $\mathcal{U} = \Re(A) + i\Re(A)$ we have $a^2 = x_o + iy_o$, where $x_o, y_o \in \Re(A)$. Then $x_o \in \Re(A)_s^+ = A^+$ and there exists a non zero spectral projection $e \in A$ for x_o and a positive number λ such that $e \leq \lambda x_o$.

Given any $t \in e\Re(A)e$ we have

$$tt^* = tet^* \leq \lambda t x_o t^* = \lambda\Phi(ta^2t^*) = \lambda\Phi((ta)(ta)^*).$$

Let $\{t_\alpha\} \subset e\Re e \subset z\mathcal{U}$, $\|t_\alpha\| \leq 1$ be an arbitrary net strongly converging to zero. By [T 1, Chap. V, Lemma 2.27] the map $x \to ax^*$ is strongly continuous on bounded subsets of $z\mathcal{U}$, since $\tau(a^2) < +\infty$. Thus $(t_\alpha a)^* = at_\alpha^* \to 0$ strongly, i.e. $(t_\alpha a)(t_\alpha a)^* \to 0$ weakly. From

$$0 \leq t_\alpha t_\alpha^* \leq \lambda\Phi((t_\alpha a \times (t_\alpha a)^*)$$

it follows that $t_\alpha t_\alpha^* \to 0$ weakly, i.e. $t_\alpha^* \to 0$ strongly. Therefore the map $t \to t^*$ is strongly continuous on the unit ball of $e\Re(A)e$.

On the other side since A is of type III eAe is not modular. From Lemma 1.3.3 the map $t \to t^*$ is strongly discontinuous on the unit ball of $\Re(eAe) \subset e\Re(A)e$. This contradiction shows that $\mathcal{U}$ is of type III.

Conversely, let $\mathcal{U}$ be a W^*-algebra of type III, and f a central projection in A such that fA is locally modular, $(\mathbf{I} - f)A$ is purely non modular. By Corollary 1.1.16 $f\mathcal{U} = \mathcal{U}(fA)$ is a semifinite W^*-algebra and hence $f = 0$, i.e. A is of type III.

Now we shall prove the theorem for JW-algebra of type II_∞. Let A be a JW-algebra of type II_∞. Then as we showed above, $\mathcal{U}$ does not contain any nonzero abelian projection. By Corollary 1.1.16 $\mathcal{U}$ is semifinite. Then again as above it does not contain any nonzero central finite projection and hence is properly infinite. Therefore $\mathcal{U}$ is a type II_∞ W^*-algebra.

Conversely, let $\mathcal{U}$ be a W^*-algebra of type II_∞, and let $A = fA + (\mathbf{I} - f)A$ the decomposition of A into the locally modular and purely non modular portions by a central projection $e \in A$. Since $(\mathbf{I} - f)\mathcal{U} = \mathcal{U}((\mathbf{I} - f)A)$ is of type III, then $\mathbf{I} - f = 0$, i.e. A is locally modular. As we mentioned above, A does not contain any non zero central modular projection, i.e. A is properly modular. Since A is reversible, then by Theorem 1.3.1 it does not contain any non zero abelian projection, i.e. A

is of type II. Therefore JW-algebra A is of type II_∞. ∎

Corollary 1.3.4. *(cf. Corollary 1.1.16) A reversible JW-algebra A is locally modular if and only if its enveloping W^*-algebra $\mathcal{U} = \mathcal{U}(A)$ is semifinite.*

Remark 1.3.5. The reversibility condition in Theorem 1.3.2 can not be omitted.

Indeed, as we showed in above, if A is an infinite dimensional spin factor, then its enveloping W^*-algebra $\mathcal{U}(A)$ can has the type I_∞, II_1, II_∞ or III, depending on its representations.

1.4. Classification of type I real and Jordan factors

Passing to the classification of Jordan factors up to isomorphism, we shall begin with study of the most simple Jordan factors - discrete Jordan factors, or Jordan factors of type I. It is known that W^*-factors of type I are described very clearly as the matrix algebras (see [N 2],[N 3]). We have almost similar situation for Jordan factor too.

We shall consider the reversible JW-factors of type I.

Theorem 1.4.1. *Each reversible JW-factor of type I_n, $n \neq 2$ belongs to one of the following classes (up to isomorphism):*

(i) $B(H)_s$ the symmetric bounded operators on a real Hilbert space H;

(ii) $B(H)_s$ the self-adjoint bounded operators on a complex Hilbert space H;

(iii) $B(H)_s$ the self-adjoint bounded operators on a quaternionian Hilbert space H.

Proof. Let A be a reversible JW-factor of type I. Then A is a JW-factor of type I_n on complex Hilbert space, and $n \geq 3$. If A is isomorphic to the self-adjoint part of a W^*-algebra, then this W^*-algebra is a W^*-factor of type I, and is isomorphic to $B(H)$ for some complex Hilbert space H, i.e. A is isomorphic to $B(H)_s$, and we have case (ii). Suppose that A is not isomorphic to the self-adjoint part of any W^*-algebra. Then A is a purely real JW-factor, i.e. $\Re(A) \bigcap i\Re(A) = \{0\}$, and then $\mathcal{U}(A) = \Re(A) + i\Re(A)$ is a W^*-factor. By Theorems 1.1.14 and 1.3.2 the W^*-factor $\mathcal{U}(A)$ is of type I and we can assume that

$$\mathcal{U}(A) \;=\; B(H) \;=\; \Re(A) + i\Re(A),$$

for some complex Hilbert space H.

Thus we can define a σ-weakly continuous mapping $\alpha : B(H) \to B(H)$, by setting

$$\alpha(x + iy) = (x - iy)^* = x^* + iy^* \quad x, y \in \Re(A)).$$

As we know from above (see Section 1.1) the map α is an involutive*-antiautomorphism of $B(H)$.

In [S 3] it is proved that there exists a conjugate linear isometry $j : H \to H$ such that

$$\alpha(x) = j^{-1}x^*j, \qquad (x \in B(H)).$$

Since A is the self-adjoint part of $\Re(A)$ we find that $x \in A$ iff x is self-adjoint and $\alpha(x) = x$, i.e.

$$A = \{x \in B(H)_s : \alpha(x) = x\} = \{x \in B(H)_s : xj = jx\}. \tag{1.5}$$

Since $\alpha^2 = id$, j^2 is a scalar multiple of the identity, say $j^2 = \lambda\mathbf{1}$, where $|\lambda| = 1$. Since j commutes with J^2, we find $j\lambda = \lambda j$. But, since j is a conjugate linear, $j\lambda = \overline{\lambda} j$. This implies $\lambda = \overline{\lambda}$, so we have $\lambda = \pm 1$.

First, assume $j^2 = \mathbf{1}$. Let $K = \{\xi \in H : j\xi = \xi\}$. Then K is a real Hilbert space, $H = K + iK$ and $j(\xi + i\eta) = \xi - i\eta$ whenever $\xi, \eta \in K$. By (1.5) $x \in A$ iff x is self-adjoint and leaves K invariant, that is, $(x \in A) \Leftrightarrow x|_K \in B(K)_s$. Since any $x \in B(H)$ is determined by its restriction to K, $A \cong B(K)_s$ follows. Thus we have case (i).

Next, assume $J^2 = -\mathbf{1}$. Define $k = ij$. It is easily verified that i, j, k satisfy the multiplication table of the unit quaternions, so H may be considered a quaternionian vector space. Also, i, j and k are isometries and skew symmetric with respect to the real part of the inner product in H. Thus H is a quaternionian Hilbert space with the inner product

$$(\xi|\eta)_H = Re(\xi|\eta) - (Re(i\xi|\eta))i - (Re(j\xi|\eta))j - (Re(k\xi|\eta))k.$$

By (1.5) the elements of A are exactly the self-adjoin $\mathbb{Q}$ - linear operators. Thus we have case (iii), and we have proved that A falls into one of the classes mentioned. ■

Having of the classification of JW-factors of type I, it is easy to obtain the theorem of classification for real factors of type I.

Theorem 1.4.2. *Each real W^*-factor of type I is contained (up to isomorphism) in one of the following two classes:*

(i) $B(H)$ *the bounded operators on a real Hilbert space* H*;*

(ii) $B(H)$ *the bounded operators on a quaternionian Hilbert space* H*.*

Proof. Let $\mathfrak{R}$ be a real W^*-factor of type I. Since $\mathfrak{R}_s = A$ is a purely real JW-factor of type I, then the proof of Theorem 1.4.2. is follows from the proof Theorem 1.4.1.

Corollary 1.4.3. *Each JW-factor of type I belongs to one of the following classes (up to isomorphism):*

(i) $B(H)_s$ *the symmetric bounded operators on a real Hilbert space* H*;*

(ii) $B(H)_s$ *the self-adjoint bounded operators on a complex Hilbert space* H*;*

(iii) $B(H)_s$ *the self-adjoint bounded operators on a quaternionian Hilbert space* H*;*

(iv) the spin factors.

Moreover, these classes are mutually disjoint, with the exceptions that the matrix algebras $M_2(\mathbf{R})_s$, $M_2(\mathbf{C})_s$ *and* $M_2(\mathbb{Q})_s$ *are all spin factors.*

Proof. Let A be a JW-factor of type I. Suppose that A isn't isomorphic to a spin factor. Then A is a JW-factor of type I_n on complex Hilbert space, $(n \geq 3)$ and in particular, A is reversible. Hence Theorem 1.4.1 implies (i), (ii) and (iii). Therefore it is sufficient to show that $M_2(\mathbf{R})_s$, $M_2(\mathbf{C})_s$ and $M_2(\mathbb{Q})_s$ are spin factors. Note that, by definition, a finite dimensional spin factor admits a basis $\mathbf{I}, s_1, ..., s_N$ where each s_m is a symmetry ($s_m^2 = \mathbf{I}$), and $s_m \circ s_n = 0$ if $m \neq n$. (Namely, let $s_1, ..., s_N$ be an orthonormal basis in H.) Defining

$$s_1 = \begin{pmatrix} 1 & 0 \\ 0 & -1 \end{pmatrix}, \quad s_2 = \begin{pmatrix} 0 & 1 \\ 1 & 0 \end{pmatrix}, \quad s_3 = \begin{pmatrix} 0 & i \\ -i & 0 \end{pmatrix},$$

$$s_4 = \begin{pmatrix} 0 & j \\ -j & 0 \end{pmatrix}, \quad s_5 = \begin{pmatrix} 0 & k \\ -k & 0 \end{pmatrix}.$$

It is checked easily that $s_m \cdot s_n = 0$ where $m \neq n$, and the spin system $S_{\mathbf{R}} = \{s_1, s_2\}$ generates $M_2(\mathbf{R})_s$, the spin system $S_{\mathbf{C}} = \{s_1, s_2, s_3\}$ generates $M_2(\mathbf{C})_s$, the spin system $S_{\mathbb{Q}} = \{s_1, s_2, s_3, s_4, s_5\}$ generates $M_2(\mathbb{Q})_s$. Therefore all these matrix algebras are spin factors.

Finally, the disjointness of the isomorphism classes, with the stated exceptions, follows by considering orthogonal minimal idempotents e, f in A and noting that $\{(e+f)A(e+f)\}$ is isomorphic to $M_2(\mathbf{R})_s$, $M_2(\mathbf{C})_s$, $M_2(\mathbb{Q})_s$, and A in the respective cases. ■

We close the Section by one auxiliary result, which we shall use in the next Section.

Lemma 1.4.4. *Let* $\mathfrak{R}$ *be a real self-adjoint algebra of operators on a complex Hilbert space* H *such that every self-adjoint operator in* $\mathfrak{R}$ *is a scalar multiple of the identity operator, i.e.* $\mathfrak{R}_s = \{\lambda \mathbf{I},\ \lambda \in \mathbf{R}\}$. *Then* $\mathfrak{R}$ *is isomorphic to either* $\mathbf{R}$, *or* $\mathbf{C}$, *or a field* $\mathbb{Q}$ *of quaternions.*

Moreover, the C^*-algebra, generated by $\mathfrak{R}$ in $B(H)$ is isomorphic to $\mathbf{C}$, *or* $\mathbf{C} \times \mathbf{C}$, *or* $M_2(\mathbf{C})$.

Proof. Let a be a nonzero operator in $\mathfrak{R}$ and suppose that there exists a sequence $\{b_n\}$ of operators in $\mathfrak{R}$ such that $b_n a \to 0$, or $ab_n \to 0$ uniformly, say $b_n a \to 0$. Then $a^* b_n^* b_n a = (b_n a)^*(b_n a) \to 0$. By hypothesis $b_n^* b_n = \alpha_n \mathbf{I}$, with $\alpha_n \in \mathbf{R}$, $n = 1, 2, \ldots$. Thus $\alpha_n a^* a \to 0$, i.e. $\alpha_n = \|b_n\|^2 \to 0$, and $\|b_n\| \to 0$. Similarly $ab_n \to 0$ implies $b_n \to 0$. Thus $\mathfrak{R}$ has no nonzero topological divisors of zero. By [K 1, Theorem 3.1] $\mathfrak{R}$ is isomorphic to one of $\mathbf{R}$, $\mathbf{C}$, or $\mathbf{Q}$. If $\mathfrak{R} \cong \mathbf{R}$ we have the first case of statement. Let $\mathcal{B}$ denote the C^*-algebra generated by $\mathfrak{R}$. Since $\mathfrak{R}$ is finite dimensional being isomorphic to $\mathbf{C}$ or $\mathbf{Q}$, $\mathcal{B} = \mathfrak{R} + i\mathfrak{R}$. Assume $\mathfrak{R} \cong \mathbf{C}$. Then the dimension of $\mathcal{B}$ as a vector space over $\mathbf{C}$ is 1 or 2. In case $dim\mathcal{B} = 1$ we have the second case of statement. Assume the dimension is 2. Then there exist two orthogonal nonzero projections e and f in $\mathcal{B}$ with $e + f = \mathbf{I}$ such that every operator in $\mathcal{B}$ is of the form $\lambda e + \mu f$ with $\lambda, \mu \in \mathbf{C}$. We may thus identify $\mathcal{B}$ with $\mathbf{C} \times \mathbf{C}$.

Let, finally, $\mathfrak{R} \cong \mathbf{Q}$. Then the dimension of $\mathfrak{R}$ as a real vector space is 4. Therefore the real dimension of $\mathcal{B}$ is less than or equal to 8, i.e. $dim\mathcal{B} \leq 4$ as a complex vector space. Since $\mathfrak{R}$ is non abelian so is $\mathcal{B}$, hence $\mathcal{B} \cong M_2(\mathbf{C})$. ■

1.5. Involutive *-antiautomorphisms of W^*-algebras

In this Section we shall describe continuous (i.e. not of type I) JW-factors. First, we note that if a JW-factor is isomorphic to the self-adjoint part of a W^*-algebra, then the problem is obviously reduced to classification of W^*-factors. So, henceforth, we restrict our attention, within the framework of the studied problem, to the case of JW-factors not isomorphic to the self-adjoint part of a W^*-algebra. Moreover, Proposition 1.1.11

implies the following property.

Proposition 1.5.1. *A purely real JW-factor A is not isomorphic to the self-adjoint part of a W^*-algebra if and only if $\mathcal{U}(A)$ is a W^*-factor.*

In the following theorem we give a general method of the construction of continuous JW-factors.

Theorem 1.5.2.

*(i) Let $\mathcal{U}$ be a continuous W^*-factor, α - an involutive *-antiautomorphism of $\mathcal{U}$. Then the set $A = \{x \in \mathcal{U}_s : \alpha(x) = x\}$ is a continuous JW-factor not isomorphic to the self-adjoint part of a W^*-algebra and the enveloping W^*-algebra $\mathcal{U}(\mathrm{A})$ of the JW-algebra A coincides with $\mathcal{U}$.*

*(ii) Conversely, given an arbitrary continuous JW-factor A, not isomorphic to the self-adjoint part of a W^*-algebra, there exists a unique involutive *-antiautomorphism α of the W^*-factor $\mathcal{U}(\mathrm{A})$ such that $A = \{x \in \mathcal{U}(A)_s : \alpha(x) = x\}$.*

Proof. Proof of the statement (i). Consider the set

$$\mathfrak{R} = \{x \in \mathcal{U} : \alpha(x) = x^*\}.$$

It is clear that $\mathfrak{R}$ is a real linear subspace of $\mathcal{U}$. If $x, y \in \mathfrak{R}$, i.e. $\alpha(x) = x^*$, $\alpha(y) = y^*$, then

$$\alpha(xy) = \alpha(y)\alpha(x) = y^*x^* = (xy)^*,$$

i.e. $xy \in \mathfrak{R}$. Moreover $\alpha(x^*) = \alpha(x)^*$, therefore $\mathfrak{R}$ is a real *-subalgebra of $\mathcal{U}$. Since the involution and every *-antiautomorphism of an arbitrary W^*-algebra are continuous in the weak topology, $\mathfrak{R}$ is weakly closed. Moreover, if $x = iy$, $x, y \in \mathfrak{R}$, then

$$x^* = \alpha(x) = i\alpha(y) = (-iy)^* = -x^*,$$

i.e. $x^* = 0$. Therefore $x = 0$, i.e. $\mathfrak{R} \bigcap i\mathfrak{R} = \{0\}$. So $\mathfrak{R}$ is a real W^*-algebra. Hence $A = \mathfrak{R}_s$ is a purely real JW-algebra, because $\mathfrak{R}(A) \subset \mathfrak{R}$, $\mathfrak{R} \bigcap i\mathfrak{R} = \{0\}$.

Further, for any $x \in \mathcal{U}$ we have:

$$x = \frac{x+\alpha(x^*)}{2} + i \cdot \frac{x-\alpha(x^*)}{2i},$$

and

$$\alpha\left(\frac{x+\alpha(x^*)}{2}\right) = \frac{\alpha(x)+x^*}{2} = \left(\frac{x+\alpha(x^*)}{2}\right)^*,$$

$$\alpha\left(\frac{x-\alpha(x^*)}{2i}\right) = \frac{\alpha(x)-x^*}{2i} = \left(\frac{x-\alpha(x^*)}{2i}\right)^*,$$

since $\alpha(\alpha(x)) = x$. Therefore

$$\frac{x+\alpha(x^*)}{2} \in \mathfrak{R}, \quad \frac{x-\alpha(x^*)}{2i} \in \mathfrak{R},$$

i.e. $\mathcal{U} = \mathfrak{R} + i\mathfrak{R}$.

Now we shall prove that A is a JW-factor. Let e be a central projection in A. For any skew-symmetric element $b \in \mathfrak{R}$ (i.e. $b^* = -b$), the element $ab - ba$ is clearly self-adjoint, i.e. $eb - be \in \mathfrak{R}_s = A$. Therefore $e(eb - be) = (eb - be)e$, i.e. $eb + be = 2ebe$. If we multiply this equality by e from the right, then we obtain $eb = ebe$. Hence $eb = be$, and thus e commutes with any skew-symmetric element from $\mathfrak{R}$. Since any element x from $\mathfrak{R}$ can be decomposed as $a + b$, where $a \in \mathfrak{R}_s = A$, $b \in \mathfrak{R}$, $b^* = -b$, we obtain that e commutes with any element from $\mathfrak{R}$. Therefore, e belongs to the centre of the algebra $\mathcal{U} = \mathfrak{R} + i\mathfrak{R}$. But $\mathcal{U}$ is a W^*-factor, therefore e $= 0$, or $\mathbf{1}$. So, any central projection of A is equal to 0, or $\mathbf{1}$, i.e. A is a purely real JW-factor.

We will now show that A contains no minimal projections (i.e., not of type I). Suppose that contrary and let q be a minimal projection in A. Then qAq is isomorphic to the field $\mathbf{R}$, i.e. $qAq = \{\lambda q,\ \lambda \in \mathbf{R}\}$. Since $\{q\mathfrak{R}q\}_s = qAq$, $q\mathfrak{R}q$ is a real W^*-algebra in which each self-adjoint operator has the form λq, where $\lambda \in \mathbf{R}$. By Lemma 1.4.4 $q\mathfrak{R}q$ is isomorphic, as a real $*$-algebra, to $\mathbf{R}$, or to $\mathbf{C}$, or to $\mathbf{Q}$. So the W^*-algebra $q\mathcal{U}q = q\mathfrak{R}q + iq\mathfrak{R}q$ is isomorphic either to $\mathbf{C}$, or to the algebra $M_2(\mathbf{C})$. In particular, $q\mathcal{U}q$ and thus also $\mathcal{U}$ contain a minimal projections, which contradicts the continuity of the W^*-factor $\mathcal{U}$. Therefore, the JW-factor A is also continuous. By Theorem 8.1 in [ASS], there exist four orthogonal projections e_1, e_2, e_3, e_4 pairwise equivalent via symmetry. This means that for all $i, j = 1, 2, 3, 4$ there exists a symmetry s_{ij} in A such that $e_j = s_{ij} e_i s_{ij}$, i.e., $e_j \in \mathfrak{R} e_i \mathfrak{R}$. So $\mathbf{1} = e_1 + e_2 + e_3 + e_4 \in \mathfrak{R} e_i \mathfrak{R}$ for each $i = \overline{1,4}$. Furthermore, $\mathfrak{R} e_i \mathfrak{R}$ is obviously a two-sided ideal in $\mathfrak{R}$. Therefore $\mathfrak{R} e_i \mathfrak{R} = \mathfrak{R}$ for all $i = \overline{1,4}$. By Martindale's theorem from the theory of Jordan algebras [Ja, Chap. 3, Theorem 6] $\mathfrak{R}$ is a perfect *-algebra, i.e., the pair $(\mathfrak{R}, \sigma)$, where σ is the embedding of $\mathfrak{R}_s$ in $\mathfrak{R}$), is the unital special universal envelope for the Jordan algebra $A = \mathfrak{R}_s$ (for details see [Ja, Chap. 2]). In particular, $\mathfrak{R}$ is algebraically generated by the Jordan algebra A. Hence $\mathfrak{R}(A) = \mathfrak{R}$ and $\mathcal{U}(A) = \mathfrak{R}(A) + i\mathfrak{R}(A) = \mathfrak{R} + i\mathfrak{R} = \mathcal{U}$. Thus A is a purely real JW-factor and $\mathcal{U}(\mathrm{A}) = \mathcal{U}$ is a W^*-factor. Therefore by Proposition 1.5.1 JW-factor A is not isomorphic to the self-adjoint part of a W^*-algebra. The proof of the statement (i) is complete.

We shall prove the statement (ii). Let A be a continuous JW-factor not isomorphic to the self-adjoint of a W^*-algebra. Since A is not of type I_2, then it is a reversible and purely real JW-factor. Then by Proposition 1.5.1 $\mathcal{U}(A)$ is a W^*-factor. Since $\mathcal{U}(A) = \mathfrak{R}(A) + i\mathfrak{R}(A)$ and $\mathfrak{R}(A) \bigcap i\mathfrak{R}(A) = \{0\}$, as we have showed in Section 1.1 the map $\alpha(a + ib) = a^* + ib^*$, $(a, b \in \mathfrak{R}(A))$ is an involutive *-antiautomorphism of the W^*-factor $\mathcal{U}(A)$, and it is evidently $A = \{x \in \mathcal{U}(A)_s : \alpha(x) = x\}$. Now we shall show the uniqueness of these *-antiautomorphism. Let β be another involutive *-antiautomorphism of $\mathcal{U}(A)$ such that

$$A = \{x \in \mathcal{U}(A)_s : \beta(x) = x\}.$$

Put $\theta = \alpha \cdot \beta$. It is clear that θ is a *-automorphism of $\mathcal{U}(A)$ and evidently θ acts on A identically. As $\mathfrak{R}(A)$ is the unital special universal envelope of A, then any Jordan automorphism of A have the unique extension to a real *-automorphism of $\mathfrak{R}(A)$. Therefore θ acts identically on $\mathfrak{R}(A)$ and thus θ acts identically on $\mathcal{U}(A) = \mathfrak{R}(A) + i\mathfrak{R}(A)$, i.e., $\theta = \alpha \cdot \beta$ is the identity map. Thus, $\beta = \alpha^{-1} = \alpha$ by involutivity of α. ∎

The proof of Theorem 1.5.2 and properties of the unital special universal envelope for Jordan algebras imply

Corollary 1.5.3. *Let A_1 and A_2 be purely real JW-factors not of type I, θ an isomorphism between A_1 and A_2. Then θ can be uniquely extended to an isomorphism of the W^*-algebras $\mathcal{U}(A_1)$ and $\mathcal{U}(A_2)$.*

Proof. In a proof of Theorem 1.5.2 we have showed that the real W^*-algebra $\Re(A_k)$ is the unital special universal envelope for Jordan algebra A_k $(k = 1, 2)$, and then the isomorphism θ between A_1 and A_2 can be uniquely extended to a real isomorphism between $\Re(A_1)$ and $\Re(A_2)$. Since $\Re(A_k) \bigcap i\Re(A_k) = \{0\}$ $(k = 1, 2)$, putting

$$\theta_2(x + iy) = \theta_1(x) + i\theta_1(y), \quad x, y \in \Re(A_1),$$

we define a complex *-isomorphism θ_2 between $\mathcal{U}(A_1) = \Re(A_1) + i\Re(A_1)$ and $\mathcal{U}(A_2) = \Re(A_2) + i\Re(A_2)$. This is the needed complex *-isomorphism. ■

This result gives us a possibility to define the notion of type III_λ, $0 \leq \lambda \leq 1$, JW-factors for JW-factors of type III not isomorphic to the self-adjoint part W^*-algebra, similarly to the Connes classification of type III factors (see [C 1]). Theorem 1.3.2 and Corollary 1.5.3 imply the correctness of the following definition.

Definition 1.5.4. Let A be a JW-factor of type III; we say that A has *type III_λ* if the W^*-factor $\mathcal{U}(A)$ has type III_λ, $0 \leq \lambda \leq 1$.

We shall give an example showing that on a Krieger factor (i.e., the crossed product of an abelian W^*-algebra and its ergodic automorphism) there always exists a canonical involutive *-antiautomorphism.

Example 1.5.5. Let N be an abelian W^*-algebra in $B(H)$, $\gamma : G \to Aut(N)$ an ergodic action of a countable discrete group G on N, and $\mathcal{U} = W^*(N, G)$ the crossed product of N and the group G acting on the Hilbert space $L_2(G, H)$ of all square integrable H-valued functions on G (see [T 1, Chap. V]). It is known that there exists a canonical isomorphism π between N and a W^*-subalgebra in $\mathcal{U}$, and $\mathcal{U}$ is generated by operators of the form $\pi(a)$ $(a \in N)$ and $u(g)$ $(g \in G)$, where

$$(\pi(a)\xi)(h) = \gamma_h^{-1}(a)\xi(h), \quad (u(g)\xi)(h) = \xi(g^{-1}h), \tag{1.6}$$

$$\xi = \xi(h) \in L_2(G, H), \; g, h \in G.$$

Each element $x \in \mathcal{U}$ can be written in the form

$$x = \sum_{g \in G} \pi(x(g))u(g),$$

where $x(\cdot)$ is an N-valued function on G. Depending on properties of N and the action of G, $\mathcal{U}$ can be a W^*-factor of any given type.

Consider the map $\alpha : \mathcal{U} \to \mathcal{U}$ given by

$$\alpha : \; x = \sum_{g \in G} \pi(x(g))u(g) \;\to\; \alpha(x) = \sum_{g \in G} \pi\Big(\gamma_g\big(x(g^{-1})\big)u(g)\Big),$$

i.e., $\alpha : x(g) \to x'(g) = \gamma_g(x(g^{-1}))$. We shall show that α is an involutive *-antiautomorphism of $\mathcal{U}$. The linearity of α is easy. Since α is transforming a function $x'(g) = \gamma_g(x(g^{-1}))$ into $\gamma_g(x'(g^{-1})) = \gamma_g\big(\gamma_g^{-1}(x(g))\big) = x(g)$, then α^2 is the identity map, i.e. α is an involutive map. It should be noted now that (1.6) for the action of the operators $\pi(a)$ and $u(g)$ implies

$$\pi\big(\gamma_g(x(g^{-1}))\big)u(g) = u(g)\pi(x(g^{-1})).$$

Therefore the action of map α can written in the form:

$$\alpha\left(\sum_{g \in G} \pi(x(g))u(g)\right) = \sum_{g \in G} u(g)\pi(x(g^{-1})),$$

and since the summation is taken over all g from group G, then in the last sum we can write g^{-1} instead of g. Finally the acting of α on $\mathcal{U}$ is written in the form

$$\alpha\left(\sum_{g \in G} \pi(x(g))u(g)\right) = \sum_{g \in G} u(g^{-1})\pi(x(g)). \tag{1.7}$$

We shall show that $\alpha(xy) = \alpha(y)\alpha(x)$, $\alpha(x^*) = \alpha(x)^*$, for all $x, y \in \mathcal{U}$. By the linearity of α it is sufficient to check these relations on the elements

$x = \pi(a)u(g)$, $y = \pi(b)u(h)$, where $a, b \in N$; $g, h \in G$. By (1.6) and (1.7) we have

$$\alpha(\pi(a)u(g)) = u(g^{-1})\pi(a) = \pi\left(\gamma_g^{-1}(a)\right) u(g^{-1});$$

similarly we have

$$\alpha(\pi(b)u(h)) = \pi\left(\gamma_g^{-1}(b)\right) u(h^{-1}).$$

Therefore according to commutativity of N, we get:

$$\alpha(\pi(b)u(h)\pi(a)u(g)) = \alpha(\pi(b)u(h)\pi(a)u(h^{-1})u(h)u(g)) =$$
$$= \alpha(\pi(b\gamma_h(a)u(hg)) = u((hg)^{-1}\pi(\gamma_h(a)b).$$

On the other hand

$$\alpha(\pi(a)u(g))\alpha(\pi(b)u(h)) = u(g^{-1})\pi(a)u(h^{-1})\pi(b) =$$
$$u(g^{-1})u(h^{-1})u(h)\pi(a)u(h^{-1})\pi(b) =$$
$$u((hg)^{-1})\pi(\gamma_h(a))\pi(b) = u((hg)^{-1})\pi(\gamma_h(a)b),$$

i.e. $\alpha(x)\alpha(y) = \alpha(yx)$. Similarly, we obtain that $\alpha(x)^* = \alpha(x^*)$. Therefore α is an involutive $*$-antiautomorphism of the W^*-factor $\mathcal{U} = W^*(N, G)$.

This example has the following generalization.

Example 1.5.6. Let $\mathcal{U} = W^*(N, G)$ be the crossed product of W^*-algebra N (not necessarily abelian) and a group G. It is known that any element from $\mathcal{U}$ can be written in the form

$$x = \sum_{g \in G} \pi(x(g))u(g).$$

If α is an $*$-antiautomorphism of N commuting with the action of the group G, then the transformation

$$\overline{\alpha}\left(\sum_{g \in G} \pi(x(g))u(g)\right) = \sum_{g \in G} u(g^{-1})\pi(\alpha(x(g))),$$

is a *-antiautomorphism of $\mathcal{U}$, and if α has period $2n$ $(n = 1, 2, \ldots)$, then $\overline{\alpha}$ has the same period. This fact is checked easily by the scheme of the Example 1.5.5, which is a particular case when N is an abelian W^*-algebra and α is the identical transformation of N. In particular, if α is an involutive *-antiautomorphism of N, then $\overline{\alpha}$ is an involutive *-antiautomorphism of $\mathcal{U} = W^*(N, G)$.

Using the example 1.5.5 we can prove the theorem of existence of continuous JW-factors of different types.

Theorem 1.5.7. *There exist JW-factors of type II_1, II_∞, III, not isomorphic to the self-adjoint part of a W^*-algebra.*

Proof. Let $\mathcal{U} = W^*(N, G)$ be a Krieger's factor considered in the Example 1.5.5. As we have noted the W^*-factor $\mathcal{U}$ can be of type II_1, II_∞ or III. Let α be the canonical involutive *-antiautomorphism of $\mathcal{U}$ constructed in 1.5.5. Put $A = \{x \in \mathcal{U}_s : \alpha(x) = x^*\}$. Then by Theorem 1.5.2 A is a continuous JW-factor not isomorphic to the self-adjoint part of any W^*-algebra, and $\mathcal{U}(A) = \mathcal{U}$. By Theorem 1.3.2 A has the same type as W^*-factor $\mathcal{U}$, i.e., A has type II_1, II_∞, III respectively . ■

Corollary 1.5.8. *For any $\lambda \in [0, 1]$ there exists a JW-factor of type III_λ not isomorphic to the self-adjoint part of a W^*-algebra.*

If we shall use the construction of *-antiautomorphism α from the Example 1.5.5, then we can give the concrete construction of JW-factors from Theorem 1.5.2 and Corollary 1.5.3. In view of importance we shall give these JW-factors as a separate example.

Example 1.5.9. Let $\mathcal{U} = W^*(N, G)$ be the W^*-factor considered in Example 1.5.5. Representing each elements of $\mathcal{U}$ in the form

$$x = \sum_{g \in G} \pi(x(g))u(g),$$

where $x(\cdot)$ is a N-valued function on G, we shall write down the concrete form of the elements of JW-factor $A = \{x \in \mathcal{U}_s : \alpha(x) = x\}$ in Theorem

1.5.2. By definition of α, the element $\alpha(x)$ is corresponding to N-valued function $\gamma_g(x(g^{-1}))$, and the element x^* is corresponding to N-valued function $\gamma_g(x(g^{-1})^*)$. Therefore elements of the real W^*-factor $\mathfrak{R} = \mathfrak{R}(A) = \{x \in \mathcal{U} : \alpha(x) = x^*\}$ in Theorem 1.5.2 correspond to such elements x that

$$\gamma_g\left(x(g^{-1})\right) = \gamma_g\left(x(g^{-1})^*\right),$$

i.e. $x(g) = x(g)^*$, for all $g \in G$. Thus elements of $\mathfrak{R}$ have the form $x = \sum_{g \in G} \pi(x(g))u(g)$, where $x(g) \in N_s$. Moreover, if $x \in A = \mathfrak{R}_s$, i.e. $x^* = x$, then $\gamma_g(x(g^{-1})) = x(g)$, for all $g \in G$. Therefore

$$\mathfrak{R} = \left\{\sum_{g \in g} \pi(x(g))u(g) \in \mathcal{U} : \ x(g) \in N_s \ \forall g \in G\right\},$$

$$A = \left\{\sum_{g \in g} \pi(x(g))u(g) \in \mathcal{U} : \ x(g) = \gamma_g(x(g^{-1})) \in N_s, \quad \forall g \in G\right\},$$

i.e. we obtain the concrete construction of JW-factors from Theorem 1.5.2 and Corollary 1.5.3.

Consider two continuous JW-factors A_1 and A_2, which are not isomorphic to the self-adjoint part of a W^*-algebra. By Theorem 1.5.2 they are defined by the couples $\left(\mathcal{U}_1, \alpha_1\right)$ and $\left(\mathcal{U}_2, \alpha_2\right)$, where $\mathcal{U}_k = \mathcal{U}(A_k)$ is the enveloping W^*-algebra for A_k, α_k is an involutive *-antiautomorphism of $\mathcal{U}_k$ such that

$$A_k = \{x \in \mathcal{U}_k : \ \alpha_k(x) = x = x^*\}, \quad k = 1, 2.$$

We find the conditions for the existence of an isomorphism between the JW-factors A_1 and A_2 in the terms of the couples $\left(\mathcal{U}_1, \alpha_1\right)$ and $\left(\mathcal{U}_2, \alpha_2\right)$.

Theorem 1.5.10. *The JW-factors A_1 and A_2 are isomorphic if and only if there exists a *-isomorphism θ between $\mathcal{U}_1$ and $\mathcal{U}_2$ such that $\alpha_2 \cdot \theta = \theta \cdot \alpha_1$.*

Proof. The proof of the "only if" part. Let A_1 and A_2 be isomorphic. Then by Corollary 1.5.3 this isomorphism has the unique extension to a *-isomorphism θ between $\mathcal{U}_1$ and $\mathcal{U}_2$. Consider the map $\alpha_1 \cdot \theta^{-1} \cdot \alpha_1 \cdot \theta$ from $\mathcal{U}_1$ into $\mathcal{U}_2$. This map evidently, is a *-automorphism of $\mathcal{U}_1$, which acts identically on A_1. Indeed, if $x \in A_1$, then $\theta(x) \in A_2$, i.e. $\alpha_2(\theta(x)) = \theta(x)$, and then $\theta^{-1} \cdot \alpha_2 \cdot \theta(x) = x$, therefore $\alpha_1 \cdot \theta^{-1} \cdot \alpha_2 \theta(x) = \alpha_1(x) = x$. By Corollary 1.5.3 the map $\alpha_1 \cdot \theta^{-1} \cdot \alpha_2 \theta$ coincides with the identity map on $\mathcal{U}_1$, i.e. $\alpha_1 \theta^{-1} \alpha_2 \theta = id$, or $\alpha_2 \theta = \theta \alpha_1$.

Proof of the "if" part. Let $\alpha_2 \cdot \theta = \theta \cdot \alpha_1$, i.e. $\alpha_2 = \theta \cdot \alpha_1 \cdot \theta^{-1}$. Then

$$A_1 = \{x \in \mathcal{U}_1 : \ x = x^* = \alpha_1(x)\},$$

$$A_2 = \{y \in \mathcal{U}_2 : \ y = y^* = \theta \cdot \alpha_1 \cdot \theta^{-1}(y)\}.$$

Hence $x \in A_1$ iff $\theta(x) \in A_2$. Therefore the restriction of θ on A_1 is an isomorphism between A_1 and A_2. ■

Corollary 1.5.11. *Let A and B be continuous JW-factors which are not isomorphic to the self-adjoint part of a W^*-algebra. Assume that they are obtained from of some W^*-factor $\mathcal{U}$ with involutive *-antiautomorphisms α and β, i.e.*

$$A = \{x \in \mathcal{U} : x = x^* = \alpha(x)\}, \qquad B = \{x \in \mathcal{U} : x = x^* = \beta(x)\}.$$

*Then A and B are isomorphic if and only if the *-antiautomorphisms α and β are conjugate, i.e. $\beta = \theta^{-1} \cdot \alpha \cdot \theta$ for some *-automorphism θ of W^*-algebra $\mathcal{U}$.*

Thus the classification of continuous JW-factors are reduced to the classification of W^*-factors and to the description of all conjugacy classes of involutive *-antiautomorphisms of W^*-factors.

By Theorem 1.4.2 any W^*-factor of type I has at most two conjugacy classes of involutive *-antiautomorphisms. In the following Section we shall give a solution of this problem for the case of injective W^*-factors of type II and III.

1.6. Involutive *-antiautomorphisms of injective W^*-algebras

The theory of injective W^*-factors (below we call them simply *injective factors*) is one of the most deep investigated parts of general W^*-algebras theory. Injective factors have interesting applications in statistical physics and quantum field theory. In this Section we shall study *-antiautomorphisms of injective factors.

For the general theory of injective factors, and in particular, the classification of injective factors we refer to works [C 4],[C 5] and monograph [T 4].

We remind that a W^*-factor $\mathcal{U}$ in $B(H)$ is said to be *injective*, if there is a projection P from $B(H)$ onto $\mathcal{U}$ with $\|P\| = 1$, $P(\mathbf{I}) = \mathbf{I}$. This is the same as $\mathcal{U}$ is approximatively finite-dimensional factor, i.e. there exists an increasing sequence $\{\mathcal{U}_n\}$ of finite-dimensional W^*-subalgebras of $\mathcal{U}$ with $\mathbf{I} \in \mathcal{U}_n$ such that $\bigcup_n \mathcal{U}_n$ is weakly dense in $\mathcal{U}$. It is known that there exists a unique (up to isomorphism) injective factor of each type II_1, II_∞, III_1, III_λ, $\lambda \neq 0$. The injective factors of type III_0 are characterized by flow of weights of these factors, that we shall give in details below.

It is known that injective factors are exactly the Krieger's factors, considered in previous Section (for injectivity of Krieger factor, see [C 4]).

In this section we shall denote the injective factor of type II_1 by R; the injective factor of type II_∞ by $R_{0,1}$ (the factor $R_{0,1}$ has the form $R_{0,1} = R \otimes B(H)$, see [C 4]); the injective factor of type III_λ, $0 \leq \lambda < 1$ by R_λ; and the injective factor of type III_1 by R_∞ (see [C 4],[H 1]).

Let $\mathcal{U}$ be a W^*-algebra with the separable predual and let $Aut(A)$ denote the group of all *-automorphisms of $\mathcal{U}$ with the topology generated by the seminorms of form

$$\theta \to \|\varphi \cdot \theta\|_*, \quad \varphi \in \mathcal{U}_*, \quad \theta \in Aut(\mathcal{U}),$$

where $\|\cdot\|_*$ is the norm of the Banach space $\mathcal{U}_*$, let $Int(\mathcal{U})$ be the group of all inner *-automorphisms of $\mathcal{U}$, $Ant(\mathcal{U})$ - the set of all *-anti-automorphisms of $\mathcal{U}$, and put $A(\mathcal{U}) = Aut(\mathcal{U}) \bigcup Ant(\mathcal{U})$.

Definition 1.6.1. We say that a W^*-factor $\mathcal{U}$ is a *McDuff factor* if it is isomorphic to the tensor product $\mathcal{U} \otimes R$.

Since the tensor product of two Krieger factors is also a Krieger factor, then the uniqueness of the Krieger factor of each type II_1, II_∞ and III_λ, $\lambda \neq 0$, implies

Proposition 1.6.2. *All Krieger factors of type II_1, II_∞ and III_λ, $\lambda \neq 0$ are McDuff factors.*

In fact, one can show that Krieger factors of type III_0 are also the McDuff factors (see [G 3]).

Definition 1.6.3. Let $\mathcal{U}$ be a W^*-algebra, H - a subgroup of the group $A(\mathcal{U})$. Two elements α and β in $A(\mathcal{U})$ are called *H-equivalent* if there exists θ in H such that $\beta = \theta \cdot \alpha \cdot \theta^{-1}$.

The main result which is necessary for classification of involutive *-antiautomorphisms of injective factors is the following theorem.

Theorem 1.6.4 [G 2],[G 3]. *Let α and β be involutive *-antiautomorphisms of a McDuff factor $\mathcal{U}$ with separable predual. Then α and β are $\overline{Int}(\mathcal{U})$-equivalent if and only if $\alpha \cdot \beta \in \overline{Int}(\mathcal{U})$.*

It is known that if R is the injective factor of type II_1, then $\overline{Int}(R) = Aut(R)$ (see [Sak 2]). Hence, Theorem 1.6.4 implies

Theorem 1.6.5. *Any two involutive *-antiautomorphisms of injective factor of type II_1 are conjugate.*

In order to consider the other cases recall the notions of smooth flow of weights and of the fundamental homomorphism for W^*-factors.

Let $\mathcal{U}$ be an infinite W^*-factor with separable $\mathcal{U}_*$, and let $W_{int}(\mathcal{U})$ be the set of all normal semifinite integrable weights on $\mathcal{U}$ (the weight φ is called *integrable* if the action of modular automorphism group σ^φ is integrable, i.e. the faithful normal operator valued weight

$$P_{\sigma^\varphi} = \int \sigma_t^\varphi(\cdot)ds \ : \mathcal{U}^+ \ \rightarrow \ (\overline{\mathcal{U}}_\varphi)^+$$

is semifinite, where $\overline{\mathcal{U}}_\varphi$ is the algebra of unbounded operators affiliated with $\mathcal{U}_\varphi$ (see [CT 2]).

Semifinite normal weights φ and ψ on $\mathcal{U}$ (may be not integrable) are said to be *equivalent* if there exists a unitary $u \in \mathcal{U}$ such that

$$uu^* = s(\varphi), \quad u^*u = s(\psi), \quad \varphi(uxu^*) = \psi(x),$$

where $s(\varphi)$, $s(\psi)$ are the supports of φ and ψ respectively , $x \in \mathcal{U}^+$.

In this case, we write $\varphi \cong \psi$ or $\psi = \varphi_u$. If there exists a unitary $u \in \mathcal{U}$ such that

$$uu^* \in \mathcal{U}_\varphi, \quad \psi(x) = \varphi(uxu^*), \ x \in \mathcal{U}^+,$$

then we write $\psi \prec \varphi$. For each projection $e \in \mathcal{U}_\varphi$ the reduced weight φ_e of φ by e (or simply subweight φ_e) is defined as

$$\varphi_e(x) = \varphi(exe), \quad x \in \mathcal{U}^+.$$

If φ and ψ have no equivalent subweights, then we say that φ and ψ are *disjoint*, and write $\varphi \perp \psi$. We say that φ and ψ are *quasiequivalent*, and write $\varphi \sim \psi$, if any non zero subweight of φ is equivalent to a subweight of ψ and any subweight of ψ is equivalent to a subweight of φ.

The quasiequivalence classes $[\varphi]$ in $W_{int}(\mathcal{U})$ form a lattice isomorphic to the lattice of all σ-finite projections of an appropriate abelian W^*-algebra $N_\mathcal{U}$. There exists a map $p_\mathcal{U}$ from the set $W_{int}(\mathcal{U})$ into the set of all projections of $N_\mathcal{U}$ and a map

$$F^\mathcal{U} : \mathbf{R}_+^* \ \rightarrow \ Aut(N_\mathcal{U})$$

such that

$$F^{\mathcal{U}}(\lambda)p_{\mathcal{U}}(\varphi) = p_{\mathcal{U}}(\lambda\varphi),$$

for all $\lambda \in \mathbf{R}_+^*$, $\varphi \in W_{int}(\mathcal{U})$, where $\mathbf{R}_+^* = (0, +\infty)$.

The couple $(N_{\mathcal{U}}, F^{\mathcal{U}})$ is called the smooth flow of weight on $\mathcal{U}$.

We denote by $Aut\left(F^{\mathcal{U}}\right)$ the group of automorphisms of $N_{\mathcal{U}}$ commuting with the image $F^{\mathcal{U}}$, i.e.

$$Aut\left(F^{\mathcal{U}}\right) = \left\{\sigma \in Aut\,(N_{\mathcal{U}}) : \sigma \cdot F^{\mathcal{U}}(\lambda) = F^{\mathcal{U}}(\lambda)\cdot\sigma \quad \text{for all } \lambda > 0\right\}.$$

For each $\alpha \in Aut(\mathcal{U})$ the transformation $\varphi \to \varphi\cdot\alpha^{-1}$ of the class of integrable weights of infinite multiplicity (see [Str]) determines a unique element $mod(\alpha)$ in $Aut\left(F^{\mathcal{U}}\right)$ such that

$$mod(\alpha)p_{\mathcal{U}}(\varphi) = p_{\mathcal{U}}\left(\varphi\cdot\alpha^{-1}\right). \tag{1.8}$$

This definition of $mod(\alpha)$ is generalized for $\alpha \in A(\mathcal{U}) = Aut(\mathcal{U}) \bigcup Ant(\mathcal{U})$ through the following result

Lemma 1.6.6 [G 2],[G 3]. *Let φ be a faithful normal semifinite weight on a W^*-algebra $\mathcal{U}$. If $\alpha \in Ant(\mathcal{U})$, then the modular automorphism group $\sigma_t^{\varphi\cdot\alpha^{-1}}$ associated with $\varphi\cdot\alpha^{-1}$ coincides with the automorphism group $\alpha\cdot\sigma_{-t}^{\varphi}\cdot\alpha^{-1}$; i.e.,*

$$\sigma_t^{\varphi\cdot\alpha^{-1}} = \alpha\cdot\sigma_{-t}^{\varphi}\cdot\alpha^{-1}, \qquad t \in \mathbf{R}.$$

Definition 1.6.7 [G 2],[G 3]. The homomorphism mod from $A(\mathcal{U})$ to $Aut\left(F^{\mathcal{U}}\right)$ definited by the formula (1.8) is called the *fundamental homomorphism.*

Proposition 1.6.8. *Let $\mathcal{U}$ be an infinite continuous Krieger factor. Then the fundamental homomorphism mod induces an isomorphism of the factor-group $Aut(\mathcal{U})/Int(\mathcal{U})$ onto the group $Aut\left(F^{\mathcal{U}}\right)$.*

Theorem 1.6.9. *Let $\mathcal{U}$ be an infinite continuous Krieger factor. Then the fundamental homomorphism mod induces a bijection between the conjugacy classes of involutive *-antiautomorphisms of $\mathcal{U}$ and the conjugacy classes of involutive automorphisms of the flow $F^{\mathcal{U}}$.*

Sketch of the proof. First, let us show that if α and β are two involutive *-antiautomorphisms of $\mathcal{U}$, then they are conjugate if and only if the involutive automorphisms $mod(\alpha)$ and $mod(\beta)$ are conjugate by some automorphism of $F^{\mathcal{U}}$. Indeed, since mod is the homomorphism, then necessity of this statement is trivial. Conversely, by the Proposition 1.6.8 the restriction of mod on $Aut(\mathcal{U})$ is a surjective map, thus we can suppose that $mod(\alpha) = mod(\beta)$, i.e. $mod(\alpha\cdot\beta) = 1$. Hence, Proposition 1.6.8 implies that $\alpha\cdot\beta \in \overline{Int}(\mathcal{U})$, and by Theorem 1.6.4 α and β are conjugate. Therefore the map mod is one-to-one between the conjugacy classes of involutive *-antiautomorphisms of $\mathcal{U}$ and the set of conjugacy classes of involutive automorphisms of $F^{\mathcal{U}}$. Moreover we obtain the map "onto". This follows from the fact that for any *-antiautomorphism α of the W^*-factor $\mathcal{U}$ such that $\alpha^2 \in \overline{Int}(\mathcal{U})$, there exists an involutive *-antiautomorphism β of the W^*-factor $\mathcal{U}$ with $mod(\alpha) = mod(\beta)$. ∎

Now we shall describe the classes of involutive *-antiautomorphisms of injective factors of type II_∞ and III_λ, $\lambda \neq 0$.

The following proposition is a generalization of Proposition IV.1.2 from [CT 2] for the case of *-antiautomorphisms.

Proposition 1.6.10. *Let $\mathcal{U}$ be a W^*-factor of type II_∞ with the separable predual. Then the map*

$$\lambda : \mathbf{R}_+^* \to F^{\mathcal{U}} \in Aut\left(F^{\mathcal{U}}\right)$$

is an isomorphism. For any $\alpha \in A(\mathcal{U})$ and each faithful normal semifinite trace τ on $\mathcal{U}$ we have:

$$\tau\cdot\alpha^{-1} = mod(\alpha)\tau,$$

where $mod(\alpha) = F^{\mathcal{U}}(\lambda)$ is identified with $\lambda \in \mathbf{R}_+^*$.

We recall that a faithful normal strictly semifinite (see Section 2.1) weight φ on W^*-factor $\mathcal{U}$ of type III_λ, $0 < \lambda < 1$ is called a *generalized trace* if $\sigma^\varphi_{T_0} = id$ for $T_0 = -\dfrac{2\pi}{\ln \lambda}$ (this exactly means that W^*-algebra $\mathcal{U}_\varphi$ is a factor, see [C 1],[CT 2]).

The following proposition is a generalization of Proposition IV.1.3 from [CT 2].

Proposition 1.6.11.

(i) If $\mathcal{U}$ be a W^*-factor of type III_λ $(0 < \lambda < 1)$ with separable predual, then the map

$$\lambda \in \mathbf{R}^*_+ \;\rightarrow\; F^{\mathcal{U}}(\lambda) \in Aut\left(F^{\mathcal{U}}\right)$$

is a homomorphism from $\mathbf{R}^*_+$ onto $Aut\left(F^{\mathcal{U}}\right)$ with the kernel $S(\mathcal{U}) \bigcap \mathbf{R}^*_+$, where $S(\mathcal{U}) = \bigcap\limits_{\varphi}\{Sp(\Delta_\varphi) : \varphi$ is a weight of $\mathcal{U}\}$. For every generalized trace φ one has

$$\varphi \cdot \alpha^{-1} \sim \lambda\varphi,$$

where $mod(\alpha) = F^{\mathcal{U}}(\lambda)$.

(ii) If $\mathcal{U}$ is a W^*-factor of type III_1, then $mod(\alpha) = 1$ for every $\alpha \in A(\mathcal{U})$.

Thus, by Proposition 1.6.10 the group of automorphisms of the smooth flow of weights on the injective type II_∞ factor $R_{0,1}$ is topologically isomorphic to the multiplicative group $\mathbf{R}^*_+$. Since $1 \in \mathbf{R}^*_+$ is the unique involutive (i.e. of period 2) element of group $\mathbf{R}^*_+$, then Theorem 1.6.9 implies

Theorem 1.6.12. *The injective factor of type II_∞ possesses exactly one conjugacy class of involutive *-antiautomorphisms .*

Let R_∞ be the Araki-Woods' factor of type III_1. Propositions 1.6.10 and 1.6.11 imply that $Int(R_\infty) = Aut(R_\infty)$, and hence, we obtain

Theorem 1.6.13. *Any two involutive *-antiautomorphisms of an injective factor of type III_1 are conjugate.*

Finally, let R_λ, $0 < \lambda < 1$ be the injective factor of type III_λ. By Proposition 1.6.11 the group of automorphisms of the smooth flow of weights is topologically isomorphic to group

$$\mathbf{R}_+^* / S(R_\lambda) \bigcap \mathbf{R}_+^* \cong \mathbf{R}_+^* / \{\lambda^n\}_{n \in Z} ,$$

This group has exactly two involutive elements, 1 and $\sqrt{\lambda}$ modulo $\{\lambda^n\}_{n\in Z}$. So Theorem 1.6.9 implies

Theorem 1.6.14. *There exist exactly two conjugacy classes on involutive *-antiautomorphisms of the injective factor of type III_λ, $0 < \lambda < 1$.*

Thus, the canonical involutive*-antiautomorphism of the Krieger factor constructed in example 1.5.5 is unique, up to conjugacy, in the case of injective factors of type II_1, II_∞, and in the Araki-Woods type III_1 factor, and also one of *-antiautomorphism of the factor R_λ $(0 < \lambda < 1)$, namely, corresponding to the element **1** in the group $\mathbf{R}_+^* / \{\lambda^n\}_{n \in Z}$.

We will construct an example of an involutive *-antiautomorphism of the W^*-factor R_λ corresponding to $\sqrt{\lambda}$ modulo $\{\lambda^n\}_{n\in Z}$.

Example 1.6.15. Suppose $\Omega = \mathbf{R} \times \mathbf{R}_+^* \subset \mathbf{R}^2$, and the measure μ on Ω is induced by the Lebesgue measure on $\mathbf{R}^2$. Let

$$G = \left\{ (a,b) = \begin{pmatrix} a & b \\ 0 & 1 \end{pmatrix} \in GL_2(Q) \mid a > 0 \right\},$$

where Q is the field of rational numbers. The action of the group G on Ω defined by

$$(a,b)(x,y) = \left(ax + b, \frac{1}{a} y \right)$$

for all $(a,b) \in G$, $(x,y) \in \Omega$, preserves the measure μ, free and ergodic. Since the group G is solvable, the crossed product $W^*(N,G)$ is isomorphic to the W^*-factor $R_{0,1}$, where $N = L^\infty(\Omega, \mu)$.

We define a one-parameter group β of automorphisms of (Ω, μ) putting

$$\beta_s(x,y) = (x, sy), \quad (x,y) \in \Omega, \; s \in \mathbf{R}_+^*.$$

Since β commutes with the action of the group G, we can extend β to a one-parameter group (also denoted by β) of automorphisms of $R_{0,1} = W^*(N,G)$. By Proposition 1.6.10 we have $mod_{R_{0,1}}(\beta_s) = s$. Let α be the canonical involutive *-antiautomorphism of $W^*(N,G)$ constructed in Example 1.5.5, i.e.

$$\alpha\left(\sum_{g\in G} \pi(a(g))u(g)\right) = \sum_{g\in G} u(g^{-1})\pi(a(g)).$$

For $s \in \mathbf{R}_+^*$ the transformation $\alpha_s = \beta_s \cdot \alpha$ is an *-antiautomorphism of $R_{0,1}$ commuting with the action β, and $mod_{R_{0,1}}(\alpha_s) = s$.

For $\lambda \in (0,1)$ consider the injective factor R_λ which is the crossed product of $R_{0,1}$ with the *-automorphism $\beta_{\lambda^{-1}}$. Let π be the canonical embedding of $R_{0,1}$ in R_λ, $v \in R_\lambda$ - a unitary such that

$$\pi(\beta_{\lambda^{-1}}(x)) = v^*\pi(x)v = Adv(\pi(x)), \quad x \in R_{0,1}.$$

Let E be the canonical conditional expectation from R_λ onto $R_{0,1}$ defined as

$$E\left(\sum_{g\in G} \pi(x(g))u(g)\right) = \pi(x(e)).$$

Consider the generalized trace Φ on R_λ defined as

$$\Phi = \tau \cdot \pi^{-1} \cdot E,$$

where τ is a faithful normal senifinite trace on $R_{0,1}$. Since for any $s \in \mathbf{R}_+^*$ the *-antiautomorphism α_s commutes with $\beta_{\lambda^{-1}}$, this implies as we noted the Example 1.5.5, that α_s uniquely extends to a *-antiautomorphism α_s of the W^*-factor R_λ such that for all

$$\sum_{n\in Z} \pi(x_n)v^* \in R_\lambda$$

we have:

$$\overline{\alpha}_s\left(\sum_{n\in Z} \pi(x_n)v^n\right) = \sum_{n\in Z} v^{-n}\pi(\alpha_s(x_n)),$$

and $mod(\overline{\alpha}_s) = s$ modulo $\{\lambda^n\}_{n \in Z}$. Indeed,

$$\Phi \cdot \overline{\alpha}_s^{-1} = \tau \cdot \pi \cdot E \cdot \overline{\alpha}_s^{-1} = \tau \cdot \overline{\alpha}_s^{-1} \cdot \pi^{-1} \cdot E = s\Phi.$$

Hence, by Proposition 1.6.11 we have that $mod(\overline{\alpha}_s) = s$.

Put $s = \lambda^{-1/2}$. Since $\alpha^2_{\lambda^{-1/2}}(x) = \beta_{\lambda^{-1}}(x)$ for all $x \in R_{0,1}$ we have

$$\overline{\alpha}^{\,2}_{\lambda^{-1/2}} = Adv,$$

because by the construction $\alpha_s(v) = v^*$ for all $s \in \mathbf{R}^*_+$. Thus there exists a unitary $\omega \in R_\lambda$, such that

$$v^* = \omega\, \overline{\alpha}^{\,2}_{\lambda^{-1/2}}(\omega^*),$$

and therefore $\gamma = Ad\omega \cdot \overline{\alpha}^{\,2}_{\lambda^{-1/2}}$ is an involutive *-antiautomorphism of R_λ with $mod(\gamma) = \sqrt{\lambda}$ modulo $\{\lambda^n\}_{n \in Z}$.

Finally we note that for injective factors of type III_0 there is no complete classification of involutive *-antiautomorphisms. It is known that for injective factors of type III_0 the flow of weights is a total invariant of isomorphism (see [C 1],[C 4]). Using Theorem 1.6.9 we can show that for each n, $1 \le n \le \infty$ there exists an injective factor of type III_0 acting in a separable Hilbert space and having exactly n conjugacy classes of involutive *-antiautomorphisms (see [G 3],[G 4]).

1.7. Classification of injective real and Jordan factors type II_1, II_∞ and III_λ, $0 < \lambda \le 1$.

Results, obtained in the preceding Sections enable us to describe the class of real and Jordan factors of type II_1, II_∞ and III_λ, $0 < \lambda \le 1$ with injective enveloping W^*-algebras.

Let A be a continuous JW-factor isomorphic to the self-adjoint part of a W^*-algebra $\mathcal{U}$. It is clear that $\mathcal{U}$ is a W^*-factor. If $\mathcal{U}$ is an injective factor then A is said to be *injective JW-factor*. Moreover, if $\mathcal{U}$ is a factor of type III_λ, $0 \leq \lambda \leq 1$, then we shall say that A is a JW-factor of type III_λ; if for a real W^*-factor $\mathfrak{R}$ its enveloping W^*-factor $\mathcal{U}(\mathfrak{R})$ has the type III_λ, $(0 \leq \lambda \leq 1)$, then we shall say that $\mathfrak{R}$ is a real W^*-factor of type III_λ (see [A 9]).

Let A be a continuous JW-factor non isomorphic to the self-adjoint part of a W^*-algebra.

Definition 1.7.1.

a) A JW-algebra A is called *injective*, if its enveloping W^*-algebra $\mathcal{U}(A)$ is injective.

b) A real W^*-algebra $\mathfrak{R}$ of $B(H)$ is called *injective*, if there exist the projection P from $B(H)$ on $\mathfrak{R}$ such that $\|P\| = 1, P(\mathbf{1}) = \mathbf{1}$.

Note that a real W^*-algebra $\mathfrak{R}$ is injective if and only if its enveloping W^*-algebra $\mathcal{U}(\mathfrak{R})$ is injective (see Proposition 2.3.8 from Chap. 2).

From the results of the provious Section and from the uniqueness of injective factor of each type II_1, II_∞, III_λ, $\lambda \neq 0$, we obtain the following theorem completely describing all injective JW-factors except JW-factors of type III_0.

Theorem 1.7.2.

(i) Up to isomorphism there exist exactly two injective JW-factors of type II_1: the first JW-factor is isomorphic to the self-adjoint part of a W^*-algebra, and the second is not isomorphic to the self-adjoint part of a W^*-algebra;

(ii) Up to isomorphism there exist exactly two injective JW-factors of type II_∞: the first JW-factor is isomorphic to the self-adjoint part of a W^*-algebra, and the second is not isomorphic to the self-adjoint part of a W^*-algebra;

(iii) Up to isomorphism there exist exactly three injective JW-factors of type $III_\lambda, 0 < \lambda < 1$: the first JW-factor is isomorphic to the self-adjoint part of a W^*-algebra, and two are not isomorphic to the self-adjoint part of a W^*-algebra;

(iv) Up to isomorphism there exist two injective JW-factors of type III_1: the first JW-factor is isomorphic to the self-adjoint part of a W^*-algebra, and the second one is not isomorphic to the self-adjoint part of a W^*-algebra.

We note that all injective JW-factors mentioned in the above theorem are constructed in the previous Sections. Namely, the injective JW-factors of type II_1, II_∞, III_λ, $0 < \lambda < 1$, and III_1, are the self-adjoint parts of the Krieger factors R_s, $(R_{0,1})_s$, $(R_\lambda)_s$, $0 < \lambda < 1$, and $(R_\infty)_s$ respectively.

The injective JW-factors of type II_1, II_∞, III_1, non isomorphic to the self-adjoint part of W^*-algebras, and one of the injective JW-factors of type III_λ, non isomorphic to the self-adjoint part of a W^*-algebra (namely, the JW-factor which is generated by an involutive *-antiautomorphism α of W^*-factor R_λ with $mod(\alpha) = 1$), was constructed in 1.5.5. Finally, the other injective JW-factor of type III_λ, non isomorphic to the self-adjoint part of a W^*-algebra - is the JW-factor $A = \{x \in R_\lambda : \ x = x^* = \gamma(x)\}$, where γ is the involutive*-antiautomorphism of R_λ with $mod(\gamma) = \sqrt{\lambda}$, from 1.6.15.

Talking about the injective factor of type III_0, it was noted in the end of Section 1.6, that we can construct a countable number of non isomorphic injective JW-factors of type III_0, with isomorphic enveloping W^*-factors.

By Theorem 1.7.2 for real injective factors we have:

Theorem 1.7.3.

(i) Up to isomorphism there exists a unique injective real W^*-factor of type II_1;

(ii) up to isomorphism there exists a unique injective real W^*-factor of type II_∞;

(iii) up to isomorphism there exist exactly two injective real W^*-factors of type III_λ, $0 < \lambda < 1$;

(iv) up to isomorphism there exists a unique injective real W^*-factor of type III_1.

Note that the real injective factors of type II_1, II_∞, III_1, and one of the real injective factors of type III_λ, are generated by involutive *-antiautomorphisms of the Krieger factors of the corresponding types. In this case if α is the *-antiautomorphism of the Krieger factor, generating these real injective factors, then $mod(\alpha) = 1$ (these real injective factors were contructed in 1.5.5). The other real injective factor of type III_λ, generated by involutive*-antiautomorphism α with $mod(\alpha) = \sqrt{\lambda}$, is constructed in 1.6.15.

For the case of real injective type III_0 factor we can construct a countable number of pairwise non isomorphic real injective factors of type III_0, with isomorphic enveloping W^*-factors (see [G 2]).

COMMENTS TO CHAPTER 1

1.1. The notion of JW-algebra was introduced firstly in the paper of Topping [To 1], devoted to the basic theory and type classification of JW-algebras. In this Section we give the classification of JW-algebras following the exposition of Ayupov [A 4], [A 9], which is slightly different from Topping's.

1.2. and 1.3. In the theory of W^*-algebras types of algebras are tightly connected with the existence of normal traces on them. In these Sections we give results on relations between the types of JW-algebras and real W^*-algebras and the existence on normal traces. Section 1.3 is devoted to study of connections between the types of JW-algebras and the types of their enveloping W^*-algebras. The main results of these Sections are due to Ayupov [A 1], [A 9], [A 10], [A 12], [A 13], except Theorem 1.3.1 which was proved by Stormer [S 2].

1.4. It is well - known that (complex) factors of type I are exactly complex matrix algebras. A similar description is obtained for real factors of type I in Theorem 1.4.2 and JW-factors of type I in Corollary 1.4.3. Theorem 1.4.1 was proved by Stormer [S 2] (see also [ASS]). Its proof is also exposed in the monograph [H-OS, Theorem 5.3.8]. Lemma 1.4.4 is due to Stormer [S 2].

1.5. Theorem 1.5.2 in this exposition was obtained by Ayupov [A 3], [A 6], [A 7], [A 8]. Different version of this theorem were given by Stormer [S 3], [S 4] and also in the monograph [H-OS, Chap. 7]. Theorem 1.5.7 was proved in [A 7], [A 8]. For the examples 1.5.5, 1.5.6 and 1.5.9 we followed the exposition of [T 1, Chap. V], [G 3], [G 4]. Theorem 1.5.10 was proved in [A 7], in more general form for JW-algebras (not only JW-

factors) it was obtained by Stormer [S 6].

1.6. This Section contains a brief exposition of the results of Giordano [G 2], [G 3], [G 4]. Theorem 1.6.5 was proved by Giordano and Jones [GJ] and independently by other methods by Stormer [S 5]. The case of injective factors of type II_∞ is considered in [St 1]. The case of general (not injective) factors is essentially more complicated. Thus Stormer has proved that given any natural number n there exists a W^*-factor of type II_1 with separable predual, which has exactly n the conjugacy classes of involutive *-antiautomorphisms [S 6, Proposition 5.2]. An analogue of Theorem 1.6.14 for σ-finite W^*-factors of type III_λ $(0 < \lambda < 1)$ was obtained by Stacey [St 4]. A description of *-antiautomorphisms of W^*-algebras (without classification up to conjugacy) was given by Stormer [S 3].

1.7. In this Section following the exposition of [A 5],[A 7],[A 8] we introduce and study injective JW-factors. As a corollary we obtain a description of injective real factors of type II_1, II_∞, and III (Theorem 1.7.3).

CHAPTER 2

REAL STRUCTURE IN W^*-ALGEBRAS

The structure theory of real W^*-algebras is relatively new, though in many aspects it is similar to the classical case of complex W^*-algebras (von Neumann algebras). In this chapter we present some result concening a deep (non elementary) structure theory of real W^*-algebras; namely the structure of type III real W^*-factors, classification of *-automorphisms and *-antiautomorphisms of real W^*-factors in comparison with the complex case, relations between the properties as injectivity, hyperfiniteness, etc.

Sections 2.1 and 2.2 are devoted to the study of type III real W^*-factors, which is based on ideas of the classical papers of Connes. Nevertheless, the discrete decomposition of real W^*-factors of type III is rather different from the complex case [C 1].

In Sections 2.3 and 2.4 we apply the technique developed by Connes and Giordano to the investigation of conjugacy of *-automorphisms and *-antiautomorphisms of real and complex injective factors.

In Section 2.5 we introduce the notions of amenability and semidiscreteness for real W^*-factors and study their relations with injectivity and hyperfiniteness. It is well known that for complex factors all these properties are equivalent [C 4], [EL], [H 1], [H 2], [W]. Unlike to the complex case for real W^*-factors the hyperfiniteness is stronger in general.

In Section 2.6 we apply the results and methods developed above to calculate diameters of the state spaces of JW-factors.

2.1. Real crossed product of real W^*-algebra by an automorphism

In this Section we obtain the discrete decomposition of real type III_λ, $0 < \lambda < 1$ factors. Discrete decomposition of complex type III factors was constructed in the fundamental work of Connes devoted to their inner structure and classification. In [C 1] he has classified them into those of type III_λ, $\lambda \in [0,1]$, and proved that the study of structure of type III_λ factors, $\lambda \neq 1$, can be reduced to the study of II_∞ W^*-algebras and outer conjugacy classes of their *-automorphisms, since any factor of type III_λ $(\lambda \neq 1)$ is isomorphic to the crossed product of some W^*-algebra of type II_∞ by an *-automorphism. Type III_1 factors doesn't admit such decomposition (see [C 1]).

We shall obtain some real analogues of this decomposition which slightly differs from the complex case and prove that each real σ-finite factor of type III_λ $(\lambda \in (0,1))$ is either a crossed product of a complex II_∞ factor by an antilinear *-automorphism or a crossed product of a real II_∞ factor by a real *-automorphism.

First we give some definitions and facts from the theory of W^*-factors. Let M be a σ-finite factor of type III, let $W_{nsf}(M)$ be a set of all normal semifinite weights on M. Studying normal weights, Connes has showed that the modular automorphism groups of all normal semifinite weights of M are connected by 1-cocycle. Namely, Connes obtained the following theorem.

Theorem. *(Radon-Nikodym type theorem of Connes for weights)*
Let $\varphi,\ \psi \in W_{nsf}(M)$. *Then there exists a unique one parameter family* $\left\{\left(\frac{\partial\psi}{\partial\varphi}\right)_t\right\}_{t\in\mathrm{R}}$ *of unitary elements from* M *such that*

(i) $$\left(\frac{\partial\psi}{\partial\varphi}\right)_{s+t} = \left(\frac{\partial\psi}{\partial\varphi}\right)_s \sigma_s^{\varphi}\left(\left(\frac{\partial\psi}{\partial\varphi}\right)_t\right);$$

(ii) $$\sigma_t^{\psi}(x) = \left(\frac{\partial\psi}{\partial\varphi}\right)_t \sigma_t^{\varphi}(x)\left(\frac{\partial\psi}{\partial\varphi}\right)_t^*, \quad x \in M$$

(iii) *for every weight* $g \in W_{nsf}(M)$ *one has:*

$$\left(\frac{\partial\psi}{\partial g}\right)_t \cdot \left(\frac{\partial g}{\partial\varphi}\right)_t = \left(\frac{\partial\psi}{\partial\varphi}\right)_t, \qquad \text{for all } t \in \mathbf{R};$$

(iv) *if* $\{u_t\}_{t\in\mathrm{R}}$ *is an ultrastrong continuous one parameter family of unitary elements from* M, *satisfying to the statement (i), then there exists a weight* $\psi \in W_{nsf}(M)$ *such that*

$$u_t = \left(\frac{\partial\psi}{\partial\varphi}\right)_t, \qquad \text{for all } t \in \mathbf{R};$$

(v) $\psi = \varphi \cdot Adu$ *for some unitary* $u \in M$ *if and only if*

$$\left(\frac{\partial\psi}{\partial\varphi}\right)_t = u^*\sigma_t^{\varphi}(u), \qquad \text{for all } t \in \mathbf{R}.$$

The family $\left\{\left(\frac{\partial\psi}{\partial\varphi}\right)_t\right\}_{t\in\mathrm{R}}$ is called the *family of Radon-Nikodym derivatives* for the weight ψ with respect to φ.

Thus the modular automorphism groups of all weights on M are weakly equivalent. Therefore we can define correctly the Connes' invariant $T(M)$:

$$T(M) = \{t \in \mathbf{R} :\ \sigma_t^{\varphi} \in Int(M) \text{ for some } \varphi \in W_{nsf}(M)\}.$$

It is clear that $T(M)$ does not depend on the choice of the weight φ. If $S(M) \neq \{0,1\}$, then invariants $S(M)$ and $T(M)$ are related, namely

$T(M)$ is the orthogonal complement of $S(M)\bigcap \mathbf{R}_+^*$ under the duality $(t,\lambda)=\lambda^{it}$.

Now let us recall the notion of the crossed product of a W^*-algebra by a locally compact topological group of its *-automorphisms. Let M be a W^*-algebra in $B(H)$, $\gamma : G \to Aut(M)$ be a group homomorphism such that each map $g \to \gamma_g$ is strongly continuous. Let $L_2(G,H)$ be the Hilbert space of all H-valued square integrable functions on G. We consider a *-algebra $\mathcal{U} \subset B(L_2(G,H))$, generated by operators of the form: $\pi_\gamma(a)(a \in M)$ and $u(g)(g \in G)$, where

$$(\pi_\gamma(a)\xi)(h) = \gamma_h^{-1}(a)\xi(h), \quad (u(g)\xi)(h) = \xi(g^{-1}h),$$

$$\xi = \xi(h) \in L_2(G,H), \; g,h \in G.$$

The algebra $\mathcal{U}$ is called *crossed product* of W^*-algebra M by G, and denoted by $W^*(M,G)$ (or $M \times_\gamma G$). Moreover, there is the canonical imbedding $\pi_\gamma : M \to \pi_\gamma(M) \subset \mathcal{U}$.

Each element $x \in \mathcal{U}$ has the form:

$$x \;=\; \sum_{g\in G} \pi_\gamma(x(g))u(g),$$

where $x(\cdot)$ is an M-valued function on G [C 1].

Note that $M \times_\gamma G$ have another description. It is known that the Hilbert space $L^2(H,G)$ is isomorphic to the Hilbert space $H \otimes L^2(G)$. Define on $L^2(G)$ an operator of right translation ρ_t as

$$(\rho_t f)(s) \;=\; \Delta^{1/2}(t)f(st),$$

where $f(s) \in L^2(G)$, and $\Delta(t)$ is the modular function of group G. Consider on $M \otimes B(L^2(\mathbf{R}))$ the one parameter automorphism group $\theta_t = \gamma_t \otimes Ad\rho_t$, $t \in G$. Then the crossed product $M \times_\gamma G$ is isomorphic to the W^*-algebra

$$\Big(M \otimes B(L^2(G))\Big)^{\theta_t} \;=\; \{x \in M \otimes B(L^2(G)) \;:\; \theta_t(x) = x, \;\; t \in G\}.$$

It is known that if γ_1 and γ_2 are different actions of the group G on M, and if they are conjugate (outer conjugate or weakly equivalent), then W^*-algebras $M \times_{\gamma_1} G$ and $M \times_{\gamma_2} G$ are isomorphic.

An important part of the theory of type III factors is the theory of discrete decomposition of type III_λ, $\lambda \neq 1$, factors. If M is a σ-finite type III_λ, $0 < \lambda < 1$, factor then exists a couple (N, θ) such that $M \cong N \times_\theta Z$, where N is a factor of type II_∞ with a faithful normal semifinite trace τ, and $\theta \in Aut(N)$ with $\tau \cdot \theta = \lambda\tau$. If M is a σ-finite factor of type III_0, then there exists a couple (N, θ) such that $M \cong N \times_\theta Z$, where N is a W^*-algebra of type II_∞ with the continuous centre and with a faithful normal semifinite trace τ, and $\theta \in Aut(N)$ with $\tau \cdot \theta \leq \lambda_0 \tau$, for some $0 < \lambda_0 < 1$, where θ act ergodically on the center of N.

Now we construct a similar theory for real factors of type III.

Let $\Re$ be a real σ-finite W^*-algebra, θ be a *-automorphism of $\Re$. For the action $\{\theta^n\}$ of the group Z on $\Re$ we introduce a real crossed product of $\Re$ by θ. Let $L^2(H, Z)$ be the Hilbert space of all integrable with square H-valued function on Z. We consider the operators on $L^2(H, Z)$ of the following form:

$$(\pi_\theta(x)\xi)(m) = \theta^{-m}(x)\xi(m);$$

$$(\lambda_\theta(n)\xi)(m) = \xi(m-n),$$

where $x \in \Re$, $n, m \in Z$, $\xi(m) \in L^2(H, Z)$. We consider the real *-algebra generated by the sets $\{\pi_\theta(\Re)\} \bigcup \{\lambda_\theta(Z)\}$. The weak closure of this *-algebra is denoted by $W_r^*(\theta, \Re)$, (or $\Re \times_\theta Z$) and said to be a real crossed product of $\Re$ by θ.

We shall describe the connection between the real crossed product $\Re$ by θ and the complex crossed product of $\mathcal{U}(\Re)$ by $\overline{\theta}$ (where $\overline{\theta}$ is the linear extension of θ onto $\mathcal{U}(\Re)$), and the extension of involutive *-anti-automorphism $\alpha \in Ant(\mathcal{U}(\Re))$, generating $\Re$, onto $\mathcal{U}(\Re) \times_{\overline{\theta}} Z$.

Proposition 2.1.1. *Let α be an involutive *-antiautomorphisms of M commuting with γ. Then there exists a unique involutive *-antiautomorphism $\tilde{\alpha}$ of $M \times_\gamma G$ such that*

(i) $\tilde{\alpha}\, \pi_\gamma = \pi_\gamma\, \tilde{\alpha}$;

(ii) $\tilde{\alpha}(\lambda(g)) = \lambda(g)^*$ *for all* $g \in G$.

Proof. Let J be the antilinear isometry of $L_2(G)$ defined by: $J(f_1 + if_2) = f_1 - if_2$, where $f_1, f_2 \in L_2(G, \mathbf{R})$, and let β be the involutive *-antiautomorphism of $B(L_2(G))$, defined by: $\beta(x) = Jx^*J$. Consider $\alpha_1 = \alpha \otimes \beta$ - an involutive *-antiautomorphism of $M \otimes B(L_2(G))$. Since

$$M \times_\gamma G \cong \left(M \otimes B(L^2(G))\right)^{\theta_t},$$

where $\theta_t = \gamma_t \otimes Ad\rho_t$ and $Ad\rho_t$ commutes with β (what we can check easily), we have that α_1 commutes with π_γ. Consider the restriction $\tilde{\alpha}$ of the *-antiautomorphism α_1 on $M \times_\gamma G$. It is easily seen to satisfy the conditions (i), (ii). ■

Proposition 2.1.2. *The *-antiautomorphism α commutes with γ if and only if each map γ_g leaves the real algebra $\mathfrak{R}$ invariant. In this case*

$$\mathfrak{R} \times_\gamma G = \{x \in M \times_\gamma G :\ \tilde{\alpha}(x) = x^*\}$$

where $\mathfrak{R} \times_\gamma G$ denotes the real crossed product of $\mathfrak{R}$ by G.

Proof. The first claim is an immediate corollary of the above results on real algebras. So suppose that α commutes with γ and $x \in \mathfrak{R} \times_\gamma G$. Proposition 2.1.1 implies that $\tilde{\alpha}(x) = x^*$. Therefore

$$\mathfrak{R} \times_\gamma G \subset \{x \in M \times_\gamma G :\ \tilde{\alpha}(x) = x^*\}.$$

Conversely, let $x \in M \times_\gamma G$ and $\tilde{\alpha}(x) = x^*$. Let P be the $*$-algebra generated by $\{\pi_\gamma(y), \lambda_\gamma(g) : y \in M, g \in G\}$. Then the element x is a weak limit of elements $\{x_n\}$ of the *-algebra P and we can assume that $\tilde{\alpha}(x_n) = x_n^*$. Since $M = \mathfrak{R} + i\mathfrak{R}$, a simple calculation shows that each element of P is of the form $a + ib$, where a, b belong to the real *-algebra generated by $\{\pi_\gamma(f), \lambda_\gamma(g) : f \in \mathfrak{R}, g \in G\}$. Then, by Proposition 2.1.1

$\tilde{\alpha}(a) = a^*$, $\tilde{\alpha}(b) = b^*$, and so $b = 0$. Hence x is an element of $\Re \times_\gamma G$. ∎

By Propositions 2.1.1 and 2.1.2 we have

Corollary 2.1.3. *Let $\Re$ be a real W^*-algebra, θ be a *-automorphism of $\Re$; α - an involutive *-antiautomorphism of the enveloping W^*-algebra $\mathcal{U}(\Re)$ with $mod(\alpha) = 1$ and let $\overline{\theta}$ be an α-invariant extension of θ onto $\mathcal{U}$. Then*

1) $\Re \times_\theta Z \, + \, i(\Re \times_\theta Z) \; = \; \mathcal{U}(\Re) \times_{\overline{\theta}} Z$;

2) $mod(\tilde{\alpha}) = 1$.

Let M be a σ-finite W^*-algebra of type $III_\lambda, 0 < \lambda < 1$. For each faithful normal semifinite weight ψ on M, let M_ψ denote the fixed point algebra of the corresponding one parameter group $\{\sigma_t^\psi : t \in \mathbf{R}\}$ of modular automorphisms, i.e.

$$M_\psi = \{x \in M : \; \sigma_t^\psi(x) = x, \; t \in \mathbf{R}\},$$

and $Z(M_\psi)$ denote the centre of M_ψ. Recall that a normal weight ψ on M_+ is said to be *strictly semifinite* if $\psi = \vee_i \omega_i$, where $\{\omega\}_{i\in I}$ is a family of the positive normal linear forms with orthogonal supports on M.

Proposition 2.1.4. *Either*

(a) there exists an α-invariant faithful strictly semifinite weight φ on M such that $\sigma_T^\varphi = id$, where $T = 2\pi/\log(\lambda)$, or

(b) there exists a α-invariant faithful strictly semifinite weight φ on M such that $\sigma_T^\varphi = Ads$, where s is a self-adjoint unitary element (a symmetry) in $Z(M_\psi)$ with $\alpha(s) = -s$.

Proof. Let ψ' be a faithful normal state on M and put $\psi = \frac{1}{2}(\psi' + \psi' \cdot \alpha)$. Then ψ is a faithful normal state. As $T = 2\pi/\ln\lambda \in T(M)$, then there exist a unitary element $u \in Z(M_\psi)$ such that $\sigma_T^\psi = Adu$. The state ψ is α-invariant and so, by Lemma 1.6.6,

$$\sigma^{\psi}_{-t}(x) = \alpha \cdot \sigma^{\psi}_t \cdot \alpha(x),$$

for all $x \in M$. From this it follows (when $t = T$) that:

$$u^* x u = \alpha(u)^* x \alpha(u),$$

therefore $\alpha(u)u^* \in M \bigcap M'$. Since M is a factor and $\alpha^2 = id$, it follows that $\alpha(u) = \pm u$. Since $\alpha(u^2) = u^2$ and $\alpha(y) = y$, for each y in the commutative W^*-algebra generated by u^2, by spectral theorem there exists positive h which satisfies $h^{-2iT} = u^2$. If $\varphi = \psi(h\cdot)$, then φ is α-invariant and, by Lemma 4.5 of [PeT], $\sigma^{\varphi}_T = Ads$, where $s = h^{iT}u$.
Since

$$h \in Z(M_\psi), \;\; u \in Z(M_\psi) \;\text{ and }\; u^2 = h^{-2iT},$$

the element s is a self-adjoint unitary element in $Z(M_\psi)$. It belongs to M_φ since

$$\sigma^{\varphi}_t(s) = h^{it} s h^{-it},$$

for all $t \in \mathbf{R}$ and $s \in Z(M_\varphi)$ because for each $x \in M_\varphi$, $x = \sigma^{\varphi}_T(x) = sxs$.

In the case when $\alpha(u) = -u$, we have $\alpha(s) = -s$ as required for the case (b). If on the other hand, $\alpha(u) = u$, then let $k = (\mathbf{I} - p) + \sqrt{\lambda} p$ where p is the central projection $p = \frac{1}{2}(\mathbf{I} - s)$ in $Z(M_\varphi)$. Regarding k as an element of the (2-dimensional) von Neumann algebra generated by s we can easily see that $k \geq 0$ and $k^{-iT} = s$. It is also easily checked that $\alpha(k) = k$, from which it follows that the state $\varphi(k\cdot)$ has the properties required for case (a) ∎

Proposition 2.1.5.

(i) In case (a) of Proposition 2.1.4, the subalgebra M_φ is a factor.

(ii) In case (b) of Proposition 2.1.4, M_φ contains a central projection p such that pM_φ and $(\mathbf{I} - p)M_\varphi$ are factors. In this case (M_φ, α) is isomorphic to $(pM_\varphi \oplus (pM_\varphi)_0, \delta)$, where $\delta(x,y) = (y,x)$ for all $x, y \in pM_\varphi$ and where $(pM_\varphi)_0$ is the opposite algebra for pM_φ.

Proof. The assertion (i) follows from the properties of the generalized trace. In the case (b) let p be the projection $\frac{1}{2}(\mathbf{1}-s) \in Z(M_\varphi)$ and let ψ be the restriction of φ to pMp. Then $\sigma_t^\psi = \sigma_t^\varphi$ on pMp and hence, for $x \in pMp$,

$$\sigma_T^\psi(x) = sxs = spxps = pxp = x.$$

Thus ψ is a generalized trace on pMp, and then $(pMp)_\psi$ is a factor. Since

$$(pMp)_\psi = pMp \bigcap M_\varphi = pM_\varphi,$$

it follows that pM_φ is a factor, and so is $(\mathbf{1}-p)M_\varphi$ by a similar argument. Finally, since $\alpha(p) = \mathbf{1} - p$, the map $\theta : a \to (pa, p\alpha(a))$ is a *-isomorphism from M_φ onto $(pM_\varphi) \oplus (pM_\varphi)_0$, such that $\theta\alpha\theta^{-1} = \delta$. ■

Proposition 2.1.6.

(i) If φ_1, φ_2 are α-invariant strictly semifinite weights on M, satisfying the conditions of the case (a) of Proposition 2.1.4, then $\left(\frac{\partial\varphi_2}{\partial\varphi_1}\right)_T = z_1\mathbf{1}$ for some complex number z_1 with $|z_1| = 1$.

(ii) If φ_1, φ_2 satisfy the case (b) of Proposition 2.1.4 then $\left(\frac{\partial\varphi_2}{\partial\varphi_1}\right)_{2T} = z_2\mathbf{1}$, for some complex number z_2 with $|z_2| = 1$.

(iii) If (i) holds, then let $z > 0$ satisfy $z^{iT} = z_1$, and if (ii) holds, then let $z > 0$ satisfy $z^{2iT} = z_2$. In both cases, there exists a unitary element u in M with $\alpha(u) = u^*$ such that $\varphi_2(x) = z\varphi_1(uxu^*)$ for all $x \in M$.

Proof. Parts (i) and (ii) follow immediately from the properties of Radon-Nikodym derivative $\left(\frac{\partial\varphi_2}{\partial\varphi_1}\right)$ for the weights φ_1 and φ_2. To prove part (iii) let $\left(\frac{\partial\varphi_2}{\partial\varphi_1}\right)_T$ be denoted by u_T for convenience. In case (i) we have $u_T = \mathbf{1}$, whereas in case (ii),

$$s_2xs_2 = u_Ts_1xs_1u_T^*,$$

for all $x \in M$, where s_1, s_2 are a self-adjoint elements constructed in Proposition 2.1.4. Hence $s_2 u_T s_1 = \mu \mathbf{I}$, for some $\mu \in \mathbf{C}$. By the choice of z,

$$\mu^2 \mathbf{I} = (\mu s_2)^2 = u_T s_1 u_T s_1 = u_{2T} = \mathbf{I},$$

and so $\mu = \pm 1$. Thus, in case (ii), $\left(\frac{\partial \varphi_2}{\partial \varphi_1}\right)_T = \pm s_2 s_1$. It follows from 1.2.2 of [C 1] that, if $\psi(x_{i,j}) = z\varphi_1(x_{11}) + \varphi_2(x_{22})$ (as in the proof of 4.3.2 of [C 1]), then in case (ii),

$$\sigma_T^{\psi}\begin{pmatrix} x_{11} & x_{12} \\ x_{21} & x_{22} \end{pmatrix} = \begin{pmatrix} \sigma_T^{z\varphi_1}(x_{11}) & \pm\sigma_T^{z\varphi_1}(x_{12})s_1 s_2 \\ \pm\sigma_T^{\varphi_2}(x_{21})s_2 s_1 & \sigma_T^{\varphi_2}(x_{22}) \end{pmatrix} =$$

$$= \begin{pmatrix} s_1 x_{11} s_1 & \pm s_1 x_{12} s_2 \\ \pm s_2 x_{21} s_1 & s_2 x_{22} s_2 \end{pmatrix} = \begin{pmatrix} s_1 & 0 \\ 0 & \pm s_2 \end{pmatrix}\begin{pmatrix} x_{11} & x_{12} \\ x_{21} & x_{22} \end{pmatrix}\begin{pmatrix} s_1 & 0 \\ 0 & \pm s_2 \end{pmatrix} =$$

$$= s\begin{pmatrix} x_{11} & x_{12} \\ x_{21} & x_{22} \end{pmatrix} s.$$

In case (i), $\sigma_T^{\psi} = id$, and in case (ii) a simple calculation shows that s belongs to $Z(M \otimes M_2(\mathbf{C}))_{\psi}$, i.e.,

$$s = \begin{pmatrix} s_1 & 0 \\ 0 & \pm s_2 \end{pmatrix} \in Z(M \otimes M_2(\mathbf{C}))_{\psi},$$

so in the both cases the strictly semifinite weight ψ on $M \otimes M_2(\mathbf{C})$ satisfies to the corresponding conditions of Proposition 2.1.4 with respect to the antiautomorphism $\tilde{\alpha}$ defined by $\tilde{\alpha}(x_{ij}) = (\alpha x_{ji})$. Then from Proposition 2.1.5 it follows that the real algebra $\mathfrak{R} = \{x \in (M \otimes M_2(\mathbf{C}))_{\psi} : \tilde{\alpha}(x) = x^*\}$ has the centre isomorphic to either $\mathbf{R}$ (in case (a)) or $\mathbf{C}$ (in case (b)). The proof of Theorem 4.3.2 of [C 1] or its real analogue then yields (using Proposition 2.1.5) a unitary element u in M with $\varphi_2(x) = z\varphi_1(uxu^*)$ and $\alpha(u) = u^*$ for all $x \in M$. ∎

Let $\mathcal{U}$ be a W^*-algebra. Recall the notion of Arveson spectrum of an action of group $\mathbf{R}$ on $\mathcal{U}$ (see [C 1],[BR]).

Let σ be a $*$-strongly continuous representation of the group $\mathbf{R}$ by $*$-automorphisms of $\mathcal{U}$. The spectrum of the element x from $\mathcal{U}$ is the set:

$$Sp_\sigma(x) = \bigcap\{z(f) : f \in L^1(\mathbf{R}), \sigma(f)x = 0\},$$

where

$$z(f) = Ker\hat{f} = \{\gamma \in \mathbf{R} : \hat{f}(\gamma) = \int_{-\infty}^{+\infty} f(t)\exp(-it\gamma)dt = 0\},$$

$\hat{f}$ is the Fourier transform of the function $f \in L^1(\mathbf{R})$, and element $\sigma(f)x$ is defined as

$$\sigma(f)x = \int_{-\infty}^{+\infty} \sigma_t(x)f(t)dt$$

where the integral is defined in ultra-weak convergence sense.

The *spectrum* of the W^*-algebra $\mathcal{U}$ under the action σ of group $\mathbf{R}$ is defined as the set

$$Sp\sigma = \bigcap\{z(f) : f \in L^1(\mathbf{R}), \sigma(f)x = 0, \text{ for every } x \in \mathcal{U}\}.$$

It is known that $Sp\sigma = \bigcup\{Sp_\gamma(x), x \in \mathcal{U}\}$ (see [C 1]).

Let e be a projection from the fixed point algebra $\mathcal{U}^\sigma$. By $Sp\sigma^e$ we shall denote the spectrum of the restriction of the action of σ on the subalgebra $e\mathcal{U}e$, and by $\mathcal{U}(\sigma, E)$ the set:

$$\mathcal{U}(\sigma, E) = \{x \in \mathcal{U} : Sp_\sigma(x) \subset E\},$$

where E is a closed subset of $\mathbf{R}$. The following Lemma gives some properties of $Sp_\sigma(\cdot)$.

Lemma 2.1.7 [Ar],[C 1].

(1) $\mathcal{U}(\sigma, E)$ *is a linear subspace of* $\mathcal{U}$,

$$\sigma_t(\mathcal{U}(\sigma, E)) = \mathcal{U}(\sigma, E) \text{ for all } t \in \mathbf{R};$$

(2) $Sp_\sigma(x^*) = -Sp_\sigma(x)$ *for all* $x \in \mathcal{U}$;

(3) let E_1, E_2 *be closed subsets of* $\mathbf{R}$, x, y *elements of* $\mathcal{U}$ *and* $Sp_\sigma(x) \subset E_1$, $Sp_\sigma(y) \subset E_2$, *then* $Sp_\sigma(x \cdot y) \subset E_1 + E_2$;

(4) $Sp_\sigma(\sigma(f)x) \subset Sp_\sigma(x) \bigcap Supp\hat{f}$, *for all* $x \in M$;

(5) let $\{V_\alpha\}_{\alpha \in A}$ *be an open covering of* $\hat{\mathbf{R}} = \mathbf{R}_+^*$. *Then there exist an index family* B *and a family* $\{g_{\alpha,\beta}\}_{\alpha \in A, \beta \in B} \subset L^1(\mathbf{R})$, *such that*

(i) $Supp\ \hat{g}_{\alpha,\beta} \subset V_\alpha$ *for all* $\alpha \in A$, $\beta \in B$;

(ii) for all $\beta \in B$ *the set of indices* $\alpha \in A$ *such that* $g_{\alpha,\beta} \neq 0$ *is a finite set* ;

(iii) $\sum_\alpha \sigma(g_{\alpha,\beta})x \rightarrow x$ *in ultrastrong topology for all* $x \in M$.

Proposition 2.1.8. *Let* φ *satisfy the condition (b) of Proposition 2.1.4 and let* $\varphi(\mathbf{I}) = \infty$. *Then there exists a unitary element* u *in* M *with* $\alpha(u) = u^*$ *such that*
(i) $(\sqrt{\lambda})\varphi(x) = \varphi(uxu^*)$ *for all* $x \in M$,

(ii) $up = (\mathbf{I} - p)u$, *where* $p = \frac{1}{2}(\mathbf{I} - s)$,

(iii) $E_\varphi(u^n) = 0$ *for each* $n \in Z \setminus \{0\}$, *where* E_φ *is the normal faithful conditional expectation from* M *onto* M_φ,

(iv) M *is generated by* M_φ *and* u.

Proof. The existence of a unitary element u with $\alpha(u) = u^*$, satisfying condition (i) immediately follows from Proposition 2.1.6 with $\varphi_1 = \varphi$, $\varphi_2 = (\sqrt{\lambda})\varphi$ and $z = 1$. Since

$$\left(\frac{\partial(\sqrt{\lambda})\varphi}{\partial\varphi}\right)_T = (\sqrt{\lambda})^{iT} = -1,$$

one has

$$\sigma_T^{\psi} = Ads', \quad \text{where} \quad s' = \begin{pmatrix} s & 0 \\ 0 & -s \end{pmatrix}.$$

Since $s' \in Z(M \otimes M_2(\mathbf{C}))_{\psi}$, it commutes with $v = u^* \otimes e_{21}$ by construction. Thus $su = -us$, from which (ii) follows.

From the properties of Radon-Nikodym derivation for weights φ and $\sqrt{\lambda}\varphi$ we obtain

$$\sigma_t^{\varphi}(u) = u \left(\frac{\partial(\sqrt{\lambda})\varphi}{\partial \varphi} \right)_t = (\sqrt{\lambda})^{iT} u,$$

for each $t \in \mathbf{R}$. Hence,

$$E_{\varphi}(u^n) = \frac{1}{2T} \int_0^{2T} \sigma_t^{\varphi}(u^n) dt = \frac{1}{2T} \int_0^{2T} \lambda^{\frac{1}{2}int} u^n dt = 0, \quad n \neq 0.$$

This establishes (iii).

Let us prove (iv). Let $\{V_n : n \in Z\}$ be an open covering of $\mathbf{R}_+^*$ such that

$$V_n \bigcap \{\lambda^{\frac{1}{2}k} : k \in Z\} = \{\lambda^{\frac{1}{2}n}\}$$

for each $n \in Z$.

Let $Sp(\sigma^{\varphi})$ be the spectrum of M under the action of modular automorphism group σ^{φ}. Put

$$M(\sigma^{\varphi}, \{\lambda^{n/2}\}) = \{x \in M : Sp_{\sigma}(x) \subset \{\lambda^{n/2}\}\}.$$

Since $\sigma_{2T}^{\varphi} = id$, it follows that $V_n \bigcap Sp(\sigma^{\varphi}) = \{\lambda^{\frac{1}{2}n}\}$ (see Lemma 2.3.8 from [C 1]). Let $f \in L_1(G)$ with $Supp(f) \subset V_n$. Thus by Lemma 2.1.7 we have

$$\sigma^{\varphi}(f)x \in M(\sigma^{\varphi}, \{\lambda^{\frac{1}{2}n}\})$$

for each $x \in M$. By Lemma 2.1.7 the union of the sets $M(\sigma^{\varphi}, \{\lambda^{\frac{1}{2}n}\})$, $n = 1, 2...$, is total in M. Since

$$\sigma_t^{\varphi}(u) = (\sqrt{\lambda})^{it} u$$

for all $t \in \mathbf{R}$, we have that $u \in M(\sigma^\varphi, \{\lambda^{\frac{1}{2}n}\})$ and hence, if $x \in M(\sigma^\varphi, \{\lambda^{\frac{1}{2}n}\})$ then, by Lemma 2.1.7 (3) $u^n x \in M_\varphi$. It follows that $x = u^{*n}(u^n x)$ belongs to the algebra generated by u and M_φ, as required. ■

Theorem 2.1.9. *Let M be a σ-finite factor of type III_λ, $0 < \lambda < 1$ and let α be an involutive *-antiautomorphism of M. Then either*

*(a) there exist a type II_∞ factor N with an α-invariant trace τ, a *-automorphism θ of N such that $\tau \cdot \theta = \lambda\tau$, and an involutive *-antiautomorphism β of N commuting with θ such that $\mathfrak{R} = (M, \alpha)$ is isomorphic to $(N \times_\theta Z, \tilde{\beta})$,*

or

*(b) there exist a type II_∞ factor N with trace τ, and a *-antiautomorphism γ of N with $\tau \cdot \gamma = (\sqrt{\lambda})\tau$, such that $\mathfrak{R} = (M, \alpha)$ is isomorphic to $((N \oplus N_0) \times_\theta Z, \tilde{\beta})$, where $\theta(x, y) = (\gamma(y), \gamma(x))$ and $\beta(x, y) = (y, x)$, for all $x, y \in N$.*

Proof. Let φ be a strictly semifinite weight on M with $\varphi(\mathbf{I}) = \infty$, and satisfying either condition (a) or condition (b) of Proposition 2.1.4. If condition (a) is satisfied then Proposition 2.1.6 can be applied with $\varphi_1 = \varphi$, $\varphi_2 = \lambda\varphi$, and $z = 1$ to obtain a unitary element u with $\alpha(u) = u^*$ such that $\lambda\varphi(x) = \varphi(uxu^*)$ for all $x \in M$. Put

$$N = M_\varphi, \quad \theta = Adu, \quad \tau = \varphi|_N, \quad \beta = \alpha|_N.$$

Then N is a factor of type II_∞ and, by Proposition 4.1.2 of [C 1], there is a *-isomorphism from M onto $N \times_\theta Z$, which maps $m \in M_\varphi$ to $\pi_\theta(m)$ and sends u to $\lambda_\theta(1)$. Since $\alpha(u) = u^*$, it is immediate from Proposition 2.1.1 that this isomorphism takes α to $\tilde{\beta}$.

If φ satisfies condition (b) of Proposition 2.1.4 then, by Proposition 2.1.8, there exists a unitary element u with $\alpha(u) = u^*$ and $(\sqrt{\lambda})\varphi(x) = \varphi(uxu^*)$ for all $x \in M$. Let $N = pM_\varphi$, where $p = \frac{1}{2}(\mathbf{I} - s)$, let τ be the restriction of φ to N, and put $\gamma(a) = u\alpha(a)u^*$ for $a \in pM_\varphi$. Since $\alpha(p) = \mathbf{I} - p$, by Proposition 2.1.8 (ii) $u\alpha(a)u^*$ belongs to pM_φ, and so it follows easily that γ is a *-antiautomorphism of pM_φ which clearly satisfies $\tau \cdot \gamma = (\sqrt{\lambda})\tau$. Let θ and β be defined as in the statement of the theorem. If $N \oplus N_0$ is identified with M_φ via the map $a \to (pa, p\alpha(a))$ then θ corresponds to Adu, and β corresponds to α. It thus suffices to show that (M, α) is isomorphic to $(M_\varphi \times_{Adu} Z, \tilde{\alpha})$. Similar to the case (a) this can be done by checking that the conditions of Proposition 4.1.2 of [C 1] are satisfied. The last easily follows from Proposition 2.1.8. ■

Corollary 2.1.10. *Let* $\mathfrak{R}$ *be a real* σ*-finite factor of type* III_λ, $0 < \lambda < 1$. *Then either*

(a) there exist a real type II_∞ *factor* F *and a *-automorphism* θ *of* F *such that* $\mathfrak{R}$ *is *-isomorphic to* $F \times_\theta Z$,

or

(b) there exist a type II_∞ *complex factor* N *and an antilinear *-automorphism* θ *of* N *such that* $\mathfrak{R}$ *is *-isomorphic to* $N \times_\theta Z$.

Proof. This is immediate from Theorem 2.1.9 and Proposition 2.1.2 (if we put $\alpha(x) = \beta(x^*)$ in case (b)).

Remark 2.1.11. In case (a) of Theorem 2.1.9, M possesses an α-invariant generalized trace and therefore $mod(\alpha) = 1$ (see Proposition 1.6.11). In case (b) let $p = \frac{1}{2}(\mathbf{1} - s)$ where $\sigma_T^\varphi = Ads$, let u be as in Proposition 2.1.8, and put $v = (\mathbf{1} - p) + pu^2$. Denote $\psi = \varphi(h\cdot)$, where $h = (\sqrt{\lambda})p + (\mathbf{1} - p)$. Since

$$h^{iT} = (\sqrt{\lambda})^{iT} p + (\mathbf{1} - p) = \mathbf{1} - 2p = s,$$

it follows that $\sigma_T^\psi = id$ so that ψ is a (non α-invariant) generalized trace. A short calculation shows that

$$\psi(\alpha(x)) = (\sqrt{\lambda})\psi(v^* x v),$$

hence $mod(\alpha) = \sqrt{\lambda}$ (see Proposition 1.6.11).

Lemma 2.1.12. *Let φ be an α-invariant faithful strictly semifinite weight on M with $\varphi(\mathbf{I}) = \infty$. Let $Z(M_\varphi)$ be isomorphic to $\mathbf{C}^2$, and let α act ergodically on $Z(M_\varphi)$. Then there exists a self-adjoint unitary element s in $Z(M_\varphi)$ with $\alpha(s) = -s$, such that $\sigma_T^\varphi = Ads$.*

Proof. Let p be a non-trivial projection in $Z(M_\varphi)$ and put $s = 2p - \mathbf{I}$. Then s is a self-adjoint unitary element such that $\alpha(s) = -s$. To complete the proof it is necessary to show that $\sigma_T^\varphi = Ads$. Put $\psi = \varphi|_{pMp}$. Then

$$(pMp)_\psi = pMp \bigcap M_\varphi = pM_\varphi$$

(see the proof of Proposition 2.1.5). Hence $(pMp)_\psi$ has trivial centre we have $\sigma_T^\psi = id$. A similar result holds for $\psi_1 = \varphi|_{(\mathbf{I}-p)M(\mathbf{I}-p)}$. Hence if u is a unitary element in $Z(M_\varphi)$ such that $\sigma_T^\varphi = Adu$ it follows that

$$u \in (pMp)' \bigcap (\mathbf{I} - p)M(\mathbf{I} - p)'.$$

Thus $pu = up = pup \in Z(pMp)$ and $(\mathbf{I} - p)u \in Z((\mathbf{I} - p)M(\mathbf{I} - p))$. It follows that $u = \lambda p + \mu(\mathbf{I} - p)$, where $\lambda, \mu \in \mathbf{C}$, and without loss of generality one may assume that $\lambda = 1$. As we showed in the proof of Proposition 2.1.4, $\alpha(u) = \pm u$ and hence $\mu = \pm 1$. If $\mu = 1$ then $u = \mathbf{I}$ and so $\sigma_T^\varphi = id$, in contrary to the choice of M_φ. Hence $\mu = -1$, and so $u = 2p - \mathbf{I} = s$, as required. ■

Lemma 2.1.13. *Let N be a factor of type II_∞, τ a faithful semifinite normal trace on N, and β a *-antiautomorphism of N with $\tau \cdot \beta = (\sqrt{\lambda})\tau$, where $0 < \lambda < 1$, and let θ be the *-automorphism $(x, y) \to (\beta(y), \beta(x))$ of $N \oplus N_0$. Then $M = (N \oplus N_0) \times_\theta Z$ is a factor of type III_λ and $\pi_\theta(N \oplus N_0)' \bigcap M \subseteq \pi_\theta(N \oplus N_0)$.*

Proof. Since θ act freely and ergodically on the centre of $(N \oplus N_0)$, thus $M = (N \oplus N_0) \times_\theta Z$ is a factor. By Theorem 3.3.1 from [C 1] the factor M is of type III_λ. Let

$$E_1 : M \to N \oplus N_0$$

be the canonical conditional expectation, let $x \in \pi_\theta(N \oplus N_0)' \bigcap M$. Then

$$yE_1(x\lambda_\theta^*(n))\lambda_\theta(n) = E_1((xy\lambda_\theta^*(n))\lambda_\theta(n) =$$

$$E_1(x\lambda_\theta^*(n) \cdot \lambda_\theta(n)y\lambda_\theta^*(n)\lambda_\theta(n) = E_1(x\lambda_\theta^*(n))\lambda_\theta(n)y.$$

Hence,

$$yE_1(x\lambda_\theta^*(n)) = E_1(x\lambda_\theta^*(n))\lambda_\theta(n)y\lambda_\theta^*(n)$$

for every $y \in \pi_\theta(N \oplus N_0)$. Since $E_1(x\lambda_\theta^*(n)) = 0$, for each $n \neq 0$, then

$$\pi_\theta(N \oplus N_0)' \bigcap M \subseteq \pi_\theta(N \oplus N_0). \quad \blacksquare$$

Theorem 2.1.14.

*(a) Let N_i be a factor of type II_∞, $i = 1,2$; α_i an involutive *-antiautomorphism of N_i, τ_i an α_i-invariant faithful semifinite normal trace on N_i, and let θ_i be a *-automorphism of N_i commuting with α_i such that $\tau_i \cdot \theta_i = \lambda\tau_i$. Then $(N_1 \times_{\theta_1} Z,\ \tilde{\alpha}_1)$ is isomorphic to $(N_2 \times_{\theta_2} Z,\ \tilde{\alpha}_2)$ if and only if there exist an isomorphism J from (N_1, α_1) onto (N_2, α_2) and a unitary element u of N_2 with $\alpha_2(u) = u^*$ such that $J \cdot \theta_1 \cdot J^{-1} \cdot \theta_2^{-1} = Adu$.*

*(b) Let N_i be a factor of type II_∞, $i = 1,2$; τ_i a faithful semifinite normal trace on N_i, and let β_i be a *-antiautomorphism of N_i with $\tau_i \cdot \beta_i = (\sqrt{\lambda})\tau_i$. Further, let α_i be the involutive *-antiautomorphism $(x,y) \to (y,x)$ of $N_i \oplus N_0$, and θ_i the *-automorphism $(x,y) \to (\beta_i(y), \beta_i(x))$ of $N_i \oplus N_0$. Then $((N_1 \oplus (N_1)_0) \times_{\theta_1} Z,\ \tilde{\alpha}_1)$ is isomorphic to $((N_2 \oplus (N_2)_0) \times_{\theta_2} Z,\ \tilde{\alpha}_2)$ if and only if there exist a (possibly antilinear) isomorphism I from N_1 onto N_2 and a unitary element u of N_2 such that $I \cdot \beta_1 \cdot I^{-1} \cdot \beta_2^{-1} = Adu$.*

Proof. (a). If there exists an isomorphism J as above then, by Proposition 3.13 of [Da], $(N_1 \times_{\theta_1} Z,\ \tilde{\alpha}_1)$ is isomorphic to $(N_2 \times_{J\theta_1 J^{-1}} Z,\ \tilde{\alpha}_2)$. Hence one may assume that $N_1 = N_2 = N$, $\alpha_1 = \alpha_2 = \alpha$ and $J = id$. We shall construct an isomorphism between $N \times_{\theta_1} Z$ and $N \times_{\theta_2} Z$ and show that this isomorphism commutes with α.

Let $M_1 = N \times_{\theta_1} Z$, $E_1 : M_1 \to N$ be the canonical conditional expectation. In the proof of Lemma 2.1.13 we showed that

$$\pi_{\theta_1}(N)' \bigcap M \subseteq \pi_{\theta_1}(N).$$

Since M is a σ-finite W^*-algebra, then there exists a faithful normal state φ on N. Let $\psi_1 = \varphi \circ E_1$. Consider the Gelfand-Naimark-Segal representation (GNS-representation) $\{H_{\psi_1}, \pi_{\psi_1}, \xi_{\psi_1}\}$ for M, constructed by ψ_1. Let $\{x_i\}_{i=-n}^n$ be elements from N, and let j_1 and j_2 be indices from $\{-n, ..., n\}$, $n \in Z_+$. Then

$$< \pi_{\psi_1}(\pi_{\theta_1}(x_{j_1})\lambda_{\theta_1}(j_1))\xi_{\psi_1},\ \pi_{\psi_1}(\pi_{\theta_1}(x_{j_2})\lambda_{\theta_1}(j_2)\xi_{\psi_1} > =$$

$$\psi_1(\lambda_{\theta_1}(-j_2)\pi_{\theta_1}(x_{j_2}^* x_{j_1})\lambda_{\theta_1}(j_1)) = \varphi(E_1(\lambda_{\theta_1}(j_1 - j_2) \cdot \pi_{\theta_1}(\theta_1^{-j_1}(x_{j_2}^* x_{j_1}))).$$

If $j_1 \neq j_2$, then $E_1(\lambda_{\theta_1}(j_1 - j_2)) = 0$, and it is easily seen that $\pi_{\psi_1}(\pi_{\theta_1}(x_j))\lambda_{\theta_1}(j)\xi_{\psi_1}$ are pairwise orthogonal vectors, and

$$\left\| \pi_{\psi_1}\left(\sum_{j=-n}^{n} \pi_{\theta_1}(x_j)\lambda_{\theta_1}(j)\xi_{\psi_1}\right)\right\|^2 = \sum_{j=-n}^{n} \varphi\left(\pi_{\theta_1}(\theta^{-j}(x_j^* x_j))\right).$$

Similarly we shall construct the GNS-representation $\{H_{\psi_2}, \pi_{\psi_2}, \xi_{\psi_2}\}$ for $\psi_2 = \varphi \circ E_2$, where, $M_2 = N \times_{\theta_2} Z$, $E_2 : M_2 \to N$ is the canonical conditional expectation. Thus we can construct the map

$$\pi_{\psi_1}\left(\sum_{j=-n}^{n} \pi_{\theta_1}(x_j)\lambda_{\theta_1}(j)\right)\xi_{\psi_1} \to \pi_{\psi_2}\left(\sum_{j=-n}^{n} \pi_{\theta_2}(x_j)\lambda_{\theta_2}(j)\right)\xi_{\psi_2},$$

which can be extended to an isometry $J : H_{\psi_1} \to H_{\psi_2}$. Since the set

$$\sum_{j=-n}^{n} \pi_{\theta_1}(x_j)\lambda_{\theta_1}(j)$$

generates M in the strong topology, and since the sets $\pi_{\psi_i}(M_i)\xi_{\psi_i}$ are total in H_{ψ_i} $(i = 1, 2)$, then J is, clearly, an isometry between H_{ψ_1} and H_{ψ_2}. Then there exists an isomorphism $J_{1,2} : M_1 \to M_2$ such that

$$\pi_{\psi_2}(J_{1,2}(x)) = J\pi_{\psi_1}J^*,$$

for all $x \in M_1$.

It is checked easily that

$$J_{1,2}(\lambda_{\theta_1}(1)) = \lambda_{\theta_2}(1),$$

and that if φ, ψ_1 and ψ_2 are the α-invariant functionals, then the isomorphism $J_{1,2}$ commutes with α. Hence,

$$(N \times_{\theta_1} Z, \tilde{\alpha}) \quad \text{and} \quad (N \times_{\theta_2} Z, \tilde{\alpha})$$

are isomorphic.

Conversely, assume for simplicity that

$$(M, \alpha) = (N_1 \times_{\theta_1} Z, \tilde{\alpha}_1) = (N_1 \times_{\theta_2} Z, \tilde{\alpha}_1).$$

Let $\varphi_i = \tau_i \cdot \pi_{\theta_i}^{-1} \cdot E_i$, $i = 1, 2$. By Proposition 2.1.6 (iii) there eixsts a unitary $v \in M$ such that $\alpha(v) = v^*$ and

$$\varphi_2(x) = z\varphi_1(vxv^*)$$

for all $x \in M$. Consider the inner *-automorphism Adv. Since $N_1 = M_{\varphi_1}$, $N_2 = M_{\varphi_2}$, we have

$$Adv(N_1) = N_2, \qquad Adv \cdot E_1 \cdot Adv^* = E_2,$$

and

$$Adv(\lambda_{\theta_1}(1))\lambda_{\theta_2}^*(1) \in \pi_{\theta_2}(N_2).$$

Then there eixsts a unitary $u \in N_2$ such that

$$Adv(\lambda_{\theta_1}(1)) = \pi_{\theta_2}(u)\lambda_{\theta_2}(1).$$

Let $J : N_1 \to N_2$ be the *-isomorphism defined by the equality:

$$Adv(\pi_{\theta_1}(x)) = \pi_{\theta_2}(J(x))$$

for all $x \in N_2$. Then it is easily seen that

$$\pi_{\theta_2}(J \cdot \theta_1(x)) = Adv(\pi_{\theta_1}(\theta_1(x))) =$$

$$Adv(\lambda_{\theta_1}(1))\pi_{\theta_1}(x)\lambda_{\theta_1}(1)^*) = Adv(\lambda_{\theta_1}(1)) \cdot Adv(\pi_{\theta_1}(x)) \cdot Adv(\lambda_{\theta_1}(1)^*) =$$

$$\pi_{\theta_2}(u)\lambda_{\theta_2}(1)\pi_{\theta_2}(J(x))\lambda_{\theta_2}(1)^*\pi_{\theta_2}(u^*) = \pi_{\theta_2}(u(\theta_2 \cdot J(x))u^*).$$

Then $J \cdot \theta_1 \cdot J^{-1} \cdot \theta_2^{-1} \in Int(N_2)$ and since

$$\alpha(\lambda_{\theta_1}(1)) = \lambda_{\theta_1}(1), \quad \alpha(\lambda_{\theta_2}(1)) = \lambda_{\theta_2}(1),$$

$\alpha(v) = v^*$, we have $\alpha(u) = u^*$.

(b). If there exists an isomorphism $I : N_1 \to N_2$ as above then either $J = I \oplus I$ or $J = \alpha \cdot * \cdot (I \oplus I)$ is an isomorphism from $N_1 \oplus (N_1)_0$ onto $N_2 \oplus (N_2)_0$ such that $J\theta_1 J^{-1}\theta_2^{-1} = Ad(u, u^*)$. Then, similar to the proof of the case (a) we construct the required isomorphism from $((N_1 \oplus (N_1)_0) \times_{\theta_1} Z, \tilde{\alpha}_1)$ onto $((N_2 \oplus (N_2)_0) \times_{\theta_2} Z, \tilde{\alpha}_2)$.

The converse can be proved by a modification of Theorem 4.1.4 of [C 1]. More precisely, let

$$(M, \alpha) = ((N_1 \oplus (N_1)_0) \times_{\theta_1} Z, \tilde{\alpha}_1) = ((N_2 \oplus (N_2)_0) \times_{\theta_2} Z, \tilde{\alpha}_2),$$

and let E_i be the conditional expectation from M onto $\pi_{\theta_i}(N_i \oplus (N_i)_0)$ $i = 1, 2$.

Put

$$\varphi_i = (\tau_i \oplus \tau_i) \cdot \pi_{\theta_i}^{-1} \cdot E_i.$$

Then, by Takesaki's Theorem on conditional expectations (see [T 3])

$$E(\sigma^{\varphi_i}(x)) = \sigma^{\varphi_i} \cdot E(x) = \sigma^{\tau_i} \cdot E(x) = E(x),$$

(since $\sigma^{\tau_i} = id$ for the traces τ_1 and τ_2). Hence we obtain:

$$\pi_{\theta_i}(N_i \oplus (N_i)_0) \subseteq M_{\varphi_i}$$

and therefore, by Lemma 2.1.13,

$$Z(M_{\varphi_i}) \subseteq Z(\pi_{\theta_i}(N_i \oplus (N_i)_0)) \cong \mathbb{C}^2.$$

If the center of M_{φ_i} is trivial then $\sigma_T^{\varphi_i} = id$. But since E_i is the canonical conditional expectation from M onto M_{φ_i} $(i = 1, 2)$, we have:

$$E_i(\lambda_{\theta_i}(1) x \lambda_{\theta_i}(1)^*) = \lambda_{\theta_i}(1) E_i(x) \lambda_{\theta_i}(1)^*,$$

for all $x \in M$ and hence

$$\varphi_i(\lambda_{\theta_i}(1) x \lambda_{\theta_i}(1)^*) = (\tau_i \oplus \tau_i) \cdot \theta_i \cdot \pi_{\theta_i}^{-1}(E_i(x)) =$$

$$= (\sqrt{\lambda})(\tau_i \oplus \tau_i) \cdot \pi_{\theta_i}^{-1}(E_i(x)) \;=\; (\sqrt{\lambda})\varphi_i(x).$$

By the properties of Radon-Nikodym derivative for the weights φ_i and φ_2 it follows that

$$\sigma_T^{\varphi_i}(\lambda_{\theta_i}(1)) \;=\; \lambda_{\theta_i}(1)\left(\frac{\partial(\sqrt{\lambda})\varphi_i}{\partial\varphi_i}\right)_T \;=\; -\lambda_{\theta_i}(1),$$

giving a contradiction. Thus $Z(M_{\varphi_i})$ is isomorphic to $\mathbb{C}^2$ and so, from Lemmas 2.1.12 and 2.1.13 and the proof of Proposition 2.1.8 (III), it follows, similarly to the proof of Theorem 4.4.2 (d) of [C 1], that

$$M_{\varphi_i} = \pi_{\theta_i}(N_i \oplus (N_i)_0).$$

Further, as in part (a), using Proposition 2.1.6 we find a unitary element w in $N_2 \oplus (N_2)_0$ such that $\alpha(w) = w^*$, and an isomorphism J from $N_1 \oplus (N_1)_0$ onto $N_2 \oplus (N_2)_0$ such that $\alpha J = J\alpha$ and $J\theta_1 J^{-1}\theta_2^{-1} = Adw$. This J is either of the form $(x, y) \to (I(x), I(y))$ for a linear isomorphism I or of the form $(x, y) \to (I(y)^*, I(x)^*)$ for an antilinear isomorphism I. In the both cases, $I\beta_1 I^{-1}\beta_2^{-1} = Adu$, where $u = (w, w^*)$, as required. ∎

Thus, by Corollary 2.1.10 and Theorem 2.1.14 we see that the structure of real factors $\mathfrak{R}$ of type $\mathrm{III}_\lambda, 0 < \lambda < 1$ is defined either by a couple (F, θ), where F is a real II_∞ factor and θ is a *-automorphism of F with $mod(\theta) = \lambda$, or by a couple (N, β), where N is a complex II_∞ factor and β is a *-antiautomorphism of N with $mod(\beta) = \sqrt{\lambda}$.

2.2. Discrete decomposition of real type III_0 factors

Let $\mathfrak{R}$ be a σ-finite real factor of type III_0, $\mathcal{U} = \mathcal{U}(\mathfrak{R})$ its enveloping type III_0 factor, and let α be the involutive *-antiautomorphism of $\mathcal{U}$, generating $\mathfrak{R}$, such that $mod(\alpha) = 1$. In this Section we will show that $\mathfrak{R}$ has a decomposition into the crossed product of some real W^*-algebra of type II_∞ with a continuous centre by a $*$-automorphism of $\mathfrak{R}$ acting ergodically on the centre of $\mathfrak{R}$.

Let φ be an α-invariant faithful normal strictly semifinite weight on $\mathcal{U}_+$. We consider the spectrum of elements of real W^*-algebra under the action of modular automorphism group $\sigma = \sigma^\varphi$ associated with φ.

Lemma 2.2.1. *Let $x \in \mathcal{U}$. Then*

$$Sp_\sigma(\alpha(x)) = -Sp_\sigma(x).$$

Proof. Let $t_0 \in Sp_\sigma(x)$. Then $\hat{f}(t_0) = 0$ for every function $f \in L^1(\mathbf{R})$, such that $\sigma(f)x = 0$. Since

$$\sigma(f)(\alpha(x)) = \int_{-\infty}^{+\infty} \sigma_t^\varphi(\alpha(x))f(t)dt,$$

and (see Lemma 1.6.6)

$$\sigma_t^\varphi = \alpha \cdot \sigma_{-t}^{\varphi\cdot\alpha} \cdot \alpha \ = \ \alpha \cdot \sigma_{-t}^\varphi \cdot \alpha,$$

and since α is ultraweak continuous (see [C 1]), we obtain

$$\int_{-\infty}^{+\infty} \alpha\left(\sigma_{-t}^\varphi(x)\right) f(t)dt \ = \ \alpha\Big(\int_{-\infty}^{+\infty} \sigma_s^\varphi(x)f(-s)ds\Big), \ \text{ where } s = -t.$$

Put $f'(t) = f(-t)$. Then

$$\hat{f}'(-t_0) \ = \ \int_{-\infty}^{+\infty} f(-t)\exp(itt_0)dt \ = \ -\hat{f}(t_0) \ = \ 0.$$

Hence $z(f) = -z(f')$, and therefore $Sp_\sigma(x) = -Sp_\sigma(\alpha(x))$. ■

Corollary 2.2.2. *Let $Sp_\sigma(x) \subset E$, where $\sigma = \sigma^\varphi$, φ an α-invariant weight on $\mathcal{U}$: suppose that E is a closed subset of $\mathbf{R}$, $x = x_1 + ix_2 \in \mathcal{U}$, $x_1, x_2 \in \mathfrak{R}$. Then*

$$Sp_\sigma(x_1) \subset E, \qquad Sp_\sigma(x_2) \subset E.$$

Proof. By Lemma 2.1.7 we have:

$$Sp_\sigma(\alpha(x^*)) = -Sp_\sigma(x^*) = Sp_\sigma(x).$$

Then by Lemma 2.2.1 we obtain: $\alpha(x^*) = (x_1 - ix_2) \in \mathcal{U}(\sigma, E)$, therefore,

$$x_1 = \frac{1}{2}(x + \alpha(x^*)) \in \mathcal{U}(\sigma, E),\ x_2 = \frac{1}{2}(x - \alpha(x^*)) \in \mathcal{U}(\sigma, E). \blacksquare$$

Now we construct on $\mathcal{U}$ an α-invariant faithful normal strictly semifinite lacunary weight following to Connes classification of complex type III_0 factors.

Definition 2.2.3 [CT 2]. A faithful normal semifinite weight φ on a W^*-algebra $\mathcal{U}$ is called *lacunary* if $\{1\}$ is an isolated point in the spectrum of the modular operator Δ_φ.

We remind that a weight φ on W^*-algebra $\mathcal{U}$ is said to be of *infinite multiplicity* if $\mathcal{U}_\varphi$ is properly infinite.

Theorem 2.2.4. *Let $\Re$ be a σ-finite real factor of type III_0, $\mathcal{U} = \mathcal{U}(\Re)$ its enveloping σ-finite type III_0 factor and let α be the involutive *-antiautomorphism of $\mathcal{U}$, generating $\Re$ and suppose that $mod(\alpha) = 1$. Then there exists a faithful normal strictly semifinite α-invariant lacunary weight φ on $\mathcal{U}_+$ of infinite multiplicity.*

In order to prove the theorem we shall give some auxiliary Lemmas.

Lemma 2.2.5. *Let $Sp\sigma \subset [-\varepsilon, \varepsilon]$, for some $\varepsilon > 0$. Then*

(1) there exists h in $\mathcal{U}^\sigma$, such that

$$-(\varepsilon/2)\mathbf{1} \le h \le (\varepsilon/2)\mathbf{1} \quad \text{and} \quad \sigma_t(x) = e^{ith} x e^{-ith},$$

for every $t \in \mathbf{R}, x \in \mathcal{U}$;

(2) If $\sigma = \sigma^\varphi$, where φ is a faithful normal strictly semifinite α-invariant weight, then

$$h \in \mathcal{U}^\sigma \bigcap \Re.$$

Proof. The statement (1) of Lemma coincides with the statement of Lemma 5.2.2 from [C 1]. For the completeness we shall give its proof. For every closed set $E \subset \mathbf{R}$ denote by $k(E)$ the projection:

$$k(E) = \bigcup\{s_l(x) : x \in \mathcal{U}, Sp_\sigma(x) \subset E\}.$$

By Theorem 3.1 from [Ar] there exists an element $b \in \mathcal{U}$ such that

$$\sigma_t(x) = e^{itb} x e^{-itb},$$

for every $t \in \mathbf{R}, x \in \mathcal{U}$, and b has the spectral decomposition

$$b = \int_{-\infty}^{+\infty} \lambda dp_\lambda,$$

where for every $\lambda \in \mathbf{R}$ the spectral projection corresponding to $[\lambda, +\infty)$, is equal to

$$\mathbf{I} - p_\lambda = \bigcap_{s<\lambda} k([s, +\infty)).$$

Then it is clear that $h = b - \frac{1}{2}\mathbf{I}$ is the needed element.

Let us prove the statement (2). It is sufficient to proof that if $\sigma = \sigma^\varphi$, and φ is an α-invariant weight, then $k([s, +\infty)) \in \mathfrak{R}$. Let $x \in \mathcal{U}$, $Sp_\sigma(x) \subset [s, +\infty)$. If $x \in \mathfrak{R}$, then $s_l(x) \in \mathfrak{R}$. Let $x = x_1 + ix_2$, where $x_1, x_2 \in \mathfrak{R}, x_i \neq 0$ and $s_l(x) = p_1 + ip_2, p_i \neq 0$. For the element $y = \alpha(x^*) = x_1 - ix_2$ it is easy to see that $Sp_\sigma(y) \subset [s, +\infty)$ and

$$s_l(y) = s_l(\alpha(x^*)) = \alpha(s_r(x^*)) = \alpha(s_l(x)) = p_1 - ip_2,$$

since $p_1^* = p_1, p_2^* = -p_2$ in view of positivity of $s_l(x)$. Note that

$$s_l(x) \bigcup \alpha(s_l(x)) = (p_1 + ip_2) \vee (p_1 - ip_2) = q$$

is a projection in $\mathfrak{R}$, because if

$$q \geq s_l(x), \qquad q \geq \alpha(s_l(x)),$$

then

$$\alpha(q) \geq \alpha(s_l(x)), \quad \alpha(q) \geq s_l(x)$$

and

$$q \wedge \alpha(q) = q = \alpha(q).$$

Thus we see that for every projection $p_1 + ip_2$ under $k([s, +\infty))$ there exists the projection $p_1 - ip_2$ less than $k([s, +\infty))$ and thus $k([s, +\infty))$ is the least upper bound of the set of projections $(p_1+ip_2) \vee (p_1 - ip_2) \in \mathfrak{R}$, that is $k([s, +\infty)) \in \mathfrak{R}$. ■

Lemma 2.2.6. *Suppose that*

$$Sp\sigma^{\varphi} \bigcap \{[-2\varepsilon, -\varepsilon] \bigcap [\varepsilon, 2\varepsilon]\} = \emptyset,$$

*for some $\varepsilon > 0$. Then exists an element $h \in \mathcal{U}_{\varphi} \bigcap \mathfrak{R}$ such that $-(\varepsilon/2)\mathbf{I} \leq h \leq (\varepsilon/2)\mathbf{I}$ and for the *-automorphism group*

$$V_t(x) = e^{-ith}\sigma_t^{\varphi}(x)e^{ith}$$

one has

$$SpV_t \bigcap (-\varepsilon, \varepsilon) = \{0\}.$$

Proof. By Lemma 2.1.7 (1), (3) it follows that the set $\mathcal{U}_1 = \mathcal{U}(\sigma^{\varphi}, [-\varepsilon, \varepsilon])$ is a W^*-subalgebra of $\mathcal{U} = \mathcal{U}(\mathfrak{R})$. By Lemma 2.1.7 one can see that $\sigma_t^{\varphi}(\mathcal{U}_1) = \mathcal{U}_1$ for every $t \in \mathbf{R}$. Moreover, Corollary 1.4.4 implies that

$$\mathcal{U}_1 = \mathcal{U}_1 \bigcap \mathfrak{R} + i\mathcal{U}_1 \bigcap \mathfrak{R}.$$

Let W be the restriction of the action of σ^{φ} on $\mathcal{U}_1$. Then $SpW \subset [-\varepsilon, \varepsilon]$. By Lemma 2.2.5 there exists an element $h \in \mathcal{U}_1 \bigcap \mathfrak{R}$ such that

$$-(1/2)\mathbf{I} \leq h \leq (1/2)\mathbf{I}, \quad W_t(h) = h,$$

for every $t \in \mathbf{R}$ and

$$W_t(x) = e^{ith}xe^{-ith}.$$

Consider the following *-automorphisms group

$$V_t(x) = e^{-ith}\sigma_t^{\varphi}(x)e^{ith},$$

where $x \in \mathcal{U}$. Then by Lemma 5.2.3 from [C 1] we obtain $SpV_t \bigcap (-\varepsilon, \varepsilon) = \{0\}$. ■

Proof of Theorem 2.2.4. Let φ_1 be a faithful normal strictly semifinite α-invariant weight on $\mathcal{U}_+$, such that

$$\alpha|_{Z(\mathcal{U}_{\varphi_1})} = id.$$

Let e_0 be a projection from $\mathfrak{R}$ such that $\varphi_1(e_0) < +\infty$ and $e_0 \neq 0$. By Lemma 2.3.4 from [C 1] there exists a projection $e \in e_0 Z(\mathcal{U}_{\varphi_1}) e_0$ $(e \neq 0)$ such that

$$Sp\sigma^e \bigcap \exp([-2\varepsilon_0, -\varepsilon_0]) \bigcap [-\varepsilon_0, 2\varepsilon_0] = \emptyset.$$

Since

$$\alpha|_{Z(\mathcal{U}_{\varphi_1})} = id,$$

we have that $e \in \mathfrak{R}$. By Lemma 2.2.6 there exists $h \in \mathcal{U}_{\varphi_1} \bigcap e\mathfrak{R}e$ such that

$$-(\varepsilon_0/2)\mathbf{I} \leq h \leq (\varepsilon_0/2)\mathbf{I}$$

and for

$$V_t(x) = e^{-ith}\sigma_t^{\varphi}(x)e^{ith}$$

one can easily see that

$$SpV_t \bigcap \exp((-\varepsilon, \varepsilon)) = \{1\}.$$

Consider the state

$$\psi(\cdot) = \varphi_{1,e}(e^{-h}\cdot).$$

For ψ we have

$$(Sp\Delta_\psi) \bigcap \exp((-\varepsilon, \varepsilon)) = \{1\}$$

and $\psi(\mathbf{I}) = 1$. Since $h \in \mathfrak{R}$, it follows that ψ is an α-invariant state:

$$\psi \cdot \alpha = \varphi_{1,e}(e^{-h}\alpha(\cdot)) = \varphi_{1,e}(e^{-h}\cdot) = \psi.$$

Since $\mathfrak{R}$ is a σ-finite real factor of type III, e and $\mathbf{I}$ are equivalent in $\mathfrak{R}$ and therefore $e\mathfrak{R}e$ is *-isomorphic to $\mathfrak{R}$. This *-isomorphism can be extended by linearity to a *-isomorphism between $e\mathcal{U}e$ and $\mathcal{U}$, and the state ψ on $e\mathcal{U}e$ is transfered to the state ψ_1 on $\mathcal{U}(\mathfrak{R})$, moreover α-invariantness of the state ψ gives α-invariantness of ψ_1, that is easy to check. The state ψ_1 is a lacunary state. Let $P_2 = \mathcal{U} \otimes F_\infty$, where $F_\infty = B(\ell^2(\mathbf{R}))$. Then there exists a *-isomorphism $\sigma_2 : P_2 \to \mathcal{U}$ such that $\sigma_2 \cdot (\alpha \otimes \alpha_2) = \alpha \cdot \sigma_2$ (where α_2 is an involutive *-antiautomorphism of F_∞). Let Tr be an α_2-invariant faithful normal semifinite trace on F_∞. Then the weight $\varphi_2 = \psi \otimes Tr$ on $(P_2)_+$ is a faithful normal strictly semifinite lacunary weight of infinite multiplicity (see [C 1, 5.3.2]), moreover, φ_2 is an $(\alpha \otimes \alpha_2)$ - invariant weight:

$$\varphi_2 \cdot (\alpha \otimes \alpha_2) = (\psi_1 \cdot \alpha) \otimes (Tr \cdot \alpha_2) = \psi_1 \cdot Tr = \varphi_2.$$

The weight $\varphi = \varphi_2 \cdot \sigma_2^{-1}$ is a faithful normal strictly semifinite lacunary weight of infinite multiplicity on $\mathcal{U}_+$, and since $\sigma_2 \cdot (\alpha \otimes \alpha_2) = \alpha \cdot \sigma_2$, the weight φ is an α-invariant weight. ■

Theorem 2.2.7. *Let Q be a σ-finite real W^*-algebra of type II_∞ with a continuous centre, θ - a *-automorphism of Q acting ergodically on the centre of Q, and let τ be a faithful normal semifinite trace on Q such that $\tau \cdot \theta \leq \lambda_0 \tau$ for a fixed λ_0, with $0 < \lambda_0 < 1$. Then the real crossed product $Q \times_\theta Z$ is a real factor of type III_0.*

Proof. Let $\mathcal{U} = \mathcal{U}(Q)$ be the enveloping W^*-algebra of Q, α an involutive *-antiautomorphism of $\mathcal{U}$ generating Q, and let $\overline{\tau}$ be the trace on $\mathcal{U}$ which is the α-invariant extension of τ on $\mathcal{U}(Q)$. We extend the *-automorphism $\theta \in Aut(Q)$ to a *-automorphism $\overline{\theta} \in Aut(\mathcal{U}(Q))$ by linearity. It is easy to see that $\overline{\theta} \cdot \alpha = \alpha \cdot \overline{\theta}$. We will show that $\overline{\tau} \cdot \overline{\theta} \leq \lambda_0 \overline{\tau}$. Let $x \in \mathcal{U}_+$. Then $(x + \alpha(x)) \in Q_+$, and we have

$$\overline{\tau}(\overline{\theta}(x)) = \frac{1}{2}\tau(\overline{\theta}(x) + \alpha(\overline{\theta}(x))) = \frac{1}{2}\tau(\overline{\theta}(x) + \overline{\theta}(\alpha(x)) =$$

$$= \frac{1}{2}\tau(\overline{\theta}(x + \alpha(x))) \leq \lambda_0 \tau(x + \alpha(x)) = \lambda_0 \overline{\tau}(x).$$

The couple $(\mathcal{U}, \overline{\theta})$ has the properties of the couples from the discrete decomposition of the type III_0 factors (see Theorem 5.1.1 from [C 1]). Indeed, $\mathcal{U}$ is a type II_∞ W^*-algebra with the continuous centre, the *-automorphism $\overline{\theta}$ act ergodically on the centre of $\mathcal{U}$ and $\tau \cdot \overline{\theta} \leq \lambda_0 \tau$ for some λ_0 $(0 < \lambda_0 < 1)$, and for the faithful normal semifinite trace τ. Therefore the crossed product $\mathcal{U} \times_{\overline{\theta}} Z$ is a factor of type III_0. By Corollary 2.1.3 (1) we have:

$$Q \times_\theta Z \; + \; i(\, Q \times_\theta Z) \; = \; \mathcal{U}(Q) \times_{\overline{\theta}} Z.$$

Therefore, $Q \times_\theta Z$ is a real factor of type III_0. ■

Theorem 2.2.8. *Let $\mathfrak{R}$ be a σ-finite real factor of type III_0, let α be the involutive *-antiautomorphism of $\mathcal{U}$ generating $\mathfrak{R}$. If $mod(\alpha) = 1$, then there exists a couple (Q, θ) satisfying the conditions of Theorem 2.2.7 and such that $\mathfrak{R}$ is *-isomorphic to $Q \times_\theta Z$.*

Proof. First, we shall outline the general scheme of the proof. By Theorem 2.2.4 there exists a faithful normal strictly semifinite α-invariant lacunary weight of infinite multiplicity on $\mathcal{U}$. In the construction of the couple (Q, θ) take Q to be equal to the real W^*-subalgebra $\mathcal{U}_\varphi \bigcap \mathfrak{R}$ (denoted by $\mathfrak{R}_\varphi$). We shall construct a unitary $u \in \mathfrak{R}$ such that

$$Adu(\mathfrak{R}_\varphi) = \mathfrak{R}_\varphi,\ Adu^*(\mathfrak{R}_\varphi) = \mathfrak{R}_\varphi,\ \ \varphi(uyu^*) \leq \lambda_0 \varphi(y),$$

for all $y \in (\mathfrak{R}_\varphi)_+$ and for some fixed λ_0, such that $0 < \lambda_0 < 1$). Moreover, $\mathfrak{R}$ is generated by $\mathfrak{R}_\varphi$ and u. Then $\mathfrak{R}$ is *-isomorphic to the real crossed product $\mathfrak{R}_\varphi$ by the *-automorphism θ, where θ is the restriction of Adu on $\mathfrak{R}_\varphi$.

Let us pass to the proof of Theorem 2.2.8. We need the following technical Lemma, which is similar to Lemmas 4.2.2 and 4.3.3 from [C 1].

Lemma 2.2.9. *Let $\mathfrak{R}$ be a real W^*-algebra, $\mathcal{U}$ the enveloping W^*-algebra of $\mathfrak{R}$, and let σ^φ be the modular *-automorphism group of $\mathcal{U}$ associated with a faithful normal strictly semifinite α-invariant lacunary weight φ on $\mathcal{U}_+$. Let $x \in \mathfrak{R}$, $s_r(x) = e$, $s_l(x) = e'$; u be a partial isometry such that $x = u|x|$. Assume that there exist projections $f, f' \in \mathcal{U}_\varphi$, such that $f \geq e, f' \geq e'$ and*

$$(Sp_\sigma(x) - Sp_\sigma(x)) \bigcap [(Sp\sigma^f) \bigcup (Sp\sigma^{f'})] = \{0\}.$$

Then $Sp_\sigma(u) = Sp_\sigma(x)$, $x^*x \in \mathfrak{R}_\varphi$ and $e, e' \in \mathfrak{R}_\varphi$. Moreover, there exist a partial isometry $v \in \mathfrak{R}$ and an element $k \in \mathfrak{R}$, such that

(1) $Sp_\sigma(v) \subset Sp_\sigma(x)$;

(2) $k \in \mathfrak{R}_\varphi$;

(3) $x = vk$;

(4) $c = v^*v \in Z(\mathfrak{R}_\varphi)$ and c is the relative central support of e in $\mathcal{U}_\varphi$;

(5) $c' = vv^* \in Z(\Re_\varphi)$ *and* c' *is the relative central support of* e' *in* $\mathcal{U}_\varphi$

(6) $v^*\mathcal{U}_\varphi v \subset \mathcal{U}_\varphi, \quad v\mathcal{U}_\varphi v^* \subset \mathcal{U}_\varphi.$

Proof of Lemma is carried out easily, similarly to the proofs of Lemmas 4.2.2 and 4.3.3 from [C 1]. But it is necessery to choose the projections and the partial isometries from the real W^*-algebra $\Re$ using Comparison theorem for JW-algebras (see [To 1]).

Consider the following sets introduced in [C 1]. Let C be the centre of $\mathcal{U}_\varphi$. Since $\varphi \cdot \alpha = \varphi$, we have

$$\mathcal{U}_\varphi = \mathcal{U}_\varphi \bigcap \Re + i\mathcal{U}_\varphi \bigcap \Re = \Re_\varphi + i\Re_\varphi.$$

Let J be the set of all partial isometries from $\mathcal{U}$ such that

(a) $v^*v \in C, \quad vv^* \in C$;

(b) $v\mathcal{U}_\varphi v^* \subset \mathcal{U}_\varphi \quad v^*\mathcal{U}_\varphi v \subset \mathcal{U}_\varphi.$

For $v_1, v_2 \in J$ put

$$v_1 \prec v_2, \; if \; e_1 = v_1^* v_1 \leq v_2^* v_2, \; and \; v_1 = v_2 e_1.$$

Let $\varepsilon_0 > 0$ be a fixed number. Define the sets E_0 and E_1 by means of the modular *-automorphism group $\sigma = \sigma^\varphi$ associated with an α-invariant weight φ:

$$E_0 = \{v \in J : Sp_\sigma(v) \subset [\varepsilon_0, +\infty), Sp_\sigma(v) - Sp_\sigma(v) \subset (-(\varepsilon_0/2), \varepsilon_0/2)\};$$

$$E_1 = \{v \in E_0 : \text{ for all } v_1, v_2 \in E_0, \text{ if } v_1 v_2 \prec v, \text{ then } v_1 v_2 = 0\}.$$

The following lemma gives some relations between the sets J, E_0, E_1 and $\Re$.

Lemma 2.2.10. *If* $\sigma^\varphi \not\equiv id$, *then*

$$J \cap \Re \neq \{0\}, \quad E_0 \cap \Re \neq \{0\} \quad E_1 \cap \Re \neq \{0\}.$$

Proof of Lemma is also carried out easily step by step, similarly to the proof of Lemma 2.1.6 in [C 1], if we apply Lemma 2.2.9 and corollary 2.2.2, and use real elements with corresponding Arveson spectrum and by constructing real partial isometries from E_0 and E_1 by the induction.

Now we shall construct the unitary $u \in \mathfrak{R}$. We consider a maximal family of elements $\{v_j\}_{j=1}^{\infty} \subset E_1 \bigcap \mathfrak{R}$ such that $v_j v_k^* = 0, v_j^* v_k = 0$ for $j \neq k$.

Lemma 2.2.11. *The element*

$$u = \sum_{j=1}^{\infty} v_j$$

is a unitary.

Proof. Let

$$u^* u = \left(\sum_{j=1}^{\infty} v_j\right)^* \left(\sum_{j=1}^{\infty} v_j\right) = e < \mathbf{1}.$$

Then in the W^*-algebra $(\mathbf{1}-e)\mathcal{U}(\mathbf{1}-e)$ the action of $\sigma_{\varphi}^{\mathbf{1}-e}$ (the restriction of σ^{φ} on $(\mathbf{1}-e)\mathcal{U}(\mathbf{1}-e)$) is non trivial. Indeed, let $\sigma_{\varphi}^{\mathbf{1}-e}$ be the trivial action. By Proposition 2.3.7 from [C 1] for every projection $f \in \mathcal{U}^{\sigma^{\varphi}}$, $f \neq 0$, there exists a representation σ_1 of $\mathfrak{R}$ on $\mathcal{U}$ such that $\sigma_1 \sim \sigma^{\varphi}$ and $Sp\sigma_1 \subset Sp\sigma^e$. Then σ^{φ} is outer conjugate to the trivial action $\sigma_{\varphi}^{\mathbf{1}-e}$. This is a contradiction. Hence, the action of $\sigma_{\varphi}^{\mathbf{1}-e}$ on $(\mathbf{1}-e)\mathcal{U}(\mathbf{1}-e)$ is non trivial. Therefore there exists an element $x' \in (\mathbf{1}-e)\mathcal{U}(\mathbf{1}-e)$, $x' \neq 0$ such that

$$Sp_{\sigma^{\mathbf{1}-e}}(x') \neq \{0\}.$$

By Lemma 2.1.7 there exists $x = x_1 + ix_2 \in (\mathbf{1}-e)\mathcal{U}(\mathbf{1}-e)$ (where $x_1, x_2 \in \mathfrak{R}$) such that

$$Sp_{\sigma^{\mathbf{1}-e}}(x) \subset (\varepsilon_0, +\infty),$$

$$Sp_{\sigma^{\mathbf{1}-e}}(x) - Sp_{\sigma^{\mathbf{1}-e}}(x) \subset [-(\varepsilon_0/2), \varepsilon_0/2].$$

Therefore, by Corollary 2.2.2,

$$Sp_{\sigma^{\mathbf{1}-e}}(x_i) \subset (\varepsilon_0, +\infty),$$

$$Sp_{\sigma^{\mathbf{1}-e}}(x_i) - Sp_{\sigma^{\mathbf{1}-e}}(x_i) \subset [-(\varepsilon_0/2), \varepsilon_0/2], \ i = 1, 2.$$

Further, by Lemma 2.2.10 there exists a partial isometry $w \in E_1 \cap \mathfrak{R}$ such that $w \neq 0$ and $w^*w \leq \mathbf{1} - e$. This contradicts the maximality of the family $\{v_j\}_{j=1}^{\infty}$, since $w^*v_j^* = v_j^*w = 0$ for all j. Therefore, $\left(\sum_{j=1}^{\infty} v_j\right)^*$ $\left(\sum_{j=1}^{\infty} v_j\right) = \mathbf{1}$. Similarly we obtain that $\left(\sum_{j=1}^{\infty} v_j\right)\left(\sum_{j=1}^{\infty} v_j\right)^* = \mathbf{1}$. ■

Now we can complete the proof of Theorem 2.2.8. By Lemmas 2.2.10 and 2.2.11 there exists a unitary $u \in \mathfrak{R}$ such that $Sp_\gamma(u) \subset (1, +\infty)$ (i.e. here $\varepsilon_0 = 1$) and

$$u\mathcal{U}_\varphi u^* = \mathcal{U}_\varphi, \qquad u^*\mathcal{U}_\varphi u = \mathcal{U}_\varphi.$$

By Lemma 5.3.4 from [C 1], u and $\mathcal{U}_\varphi$ generate $\mathcal{U}$, and it is easy to see that they satisfy the conditions of Lemma 4.1.2 from [C 1]. Thus $\mathcal{U}$ is *-isomorphic to $\mathcal{U}_\varphi \times_{\overline{\theta}} Z$, where $\overline{\theta} = Adu|_{\mathcal{U}_\varphi}$ is the restriction of Adu on $\mathcal{U}_\varphi$.

Let $\mathfrak{R}_1$ be a real W^*-subalgebra of $\mathfrak{R}$ generated by $\mathfrak{R}_\varphi$ and u. Since $\mathcal{U}_\varphi$ and u are generating $\mathcal{U}$, then $\mathcal{U}(\mathfrak{R}_1) = \mathcal{U}(\mathfrak{R})$ and $\mathfrak{R}_1 = \mathfrak{R}$. By Lemma 4.1.2 from [C 1] there exists a *-isomorphism between $\mathcal{U}$ and $\mathcal{U}_\varphi \times_{\overline{\theta}} Z$ which according to Proposition 2.1.1 is an isomorphism of $\mathfrak{R}$ onto $\mathfrak{R}_\varphi \times_\theta Z$, where $\theta = Adu|_{\mathfrak{R}_\varphi}$. By Theorem 5.3.1 from [C 1], we have: $\varphi(uyu^*) \leq \lambda_0\varphi(y)$ for some fixed λ_0, (where $0 < \lambda_0 < 1$) and for every $y \in \mathcal{U}_\varphi$. Thus $\varphi \cdot \theta \leq \lambda_0\varphi$. ■

Corollary 2.1.3 implies the converse statement of the Theorem 2.2.8.

Corollary 2.2.12. *If $\mathfrak{R}$ is isomorphic to $Q \times_\theta Z$ for an appropriate couple (Q, θ), then $mod(\alpha) = 1$ for the involutive *-antiautomorphism $\alpha \in Ant(\mathcal{U}(\mathfrak{R}))$ generating $\mathfrak{R}$.*

First we find conditions for isomorphism of real factors of type III_0. Let $\mathfrak{R}$ be a real W^*-algebra with a continuous centre, let $\mathcal{U} = \mathcal{U}(\mathfrak{R})$ be the enveloping W^*-algebra of $\mathfrak{R}$, let θ be a *-automorphism of $\mathfrak{R}$, and $\overline{\theta} \in Aut(\mathcal{U})$ the linear extension of θ on $\mathcal{U}$. Let $\overline{\theta}$ act ergodically on the centre of $\mathcal{U}$. We consider the *-automorphism $\overline{\theta}_1$ (the restriction of $\overline{\theta}$ on $Z(\mathcal{U})$). By [Kr, see p. 88, §4] for every nonzero projection $e \in Z(\mathcal{U})$, $(e \neq \mathbf{1})$ there exists a unique sequence of mutually orthogonal projections

$\{e_n\}_{n=1}^{\infty} \subset Z(\mathcal{U})$ such that

(1) $\sum\limits_{n=1}^{\infty} e_n = e$;

(2) $e\overline{\theta}_1{}^m(e_n) = 0$, if $m < n$;

(3) $\overline{\theta}_1{}^n(e_n) < e$ and $\sum\limits_{n=1}^{\infty} \overline{\theta}_1{}^n(e_n) = e$

(see. [C 1, p. 240]).

Consider the *-automorphism $\overline{\theta}_e$ on the W^*-algebra $e\mathcal{U}$

$$\overline{\theta}_e(x) = \sum_{n=1}^{\infty} \overline{\theta}^{\,n}(xe_n), \quad \text{where} \quad x \in e\mathcal{U}.$$

Lemma 2.2.13. *Let e be a projection from $Z(\mathfrak{R})$ such that $e \neq 0, e \neq \mathbf{I}$, and $\{e_n\}_{n=1}^{\infty}$ the corresponding sequence of projections in $Z(\mathcal{U})$. Then*

$$\{e_n\}_{n=1}^{\infty} \subset Z(\mathfrak{R}).$$

Proof. Suppose that there exists $e_i \in \{e_n\}_{n=1}^{\infty}$ such that $e_i \notin \mathfrak{R}$. Let α be the involutive *-antiautomorphism of $\mathcal{U}$, generating $\mathfrak{R}$. Then $\alpha(e_i) \neq e_i$. But the sequence $\{\alpha(e_n)\}_{n=1}^{\infty}$ has the same properties:

(1) $\sum\limits_{n=1}^{\infty} \alpha(e_n) = \alpha\left(\sum\limits_{n=1}^{\infty} e_n\right) = \alpha(e) = e$;

(2) $e\overline{\theta}_1{}^m(\alpha(e_n)) = \alpha\left(e\overline{\theta}_1{}^m(e_n)\right) = 0, \qquad m < n$;

(3) $\overline{\theta}_1{}^n(\alpha(e_n)) = \alpha\left(\overline{\theta}_1{}^n(e_n)\right) < e,$

since $\overline{\theta}_1{}^n$ commutes with α by the construction. Since $\{e_n\}_{n=1}^{\infty}$ is uniquely determined, we obtain a contradiction. Hence, $\alpha(e_i) = e_i$ for all i. ■

By the Lemma 2.2.13 we can introduce the following *-automorphism θ_e on $e\mathfrak{R}$:

$$\theta_e(x) = \sum_{n=1}^{\infty} \theta^n(xe_n), \quad \text{where} \quad x \in e\mathfrak{R},$$

and the linear extension of θ_e on $e\mathcal{U}$ coincides with $\overline{\theta}_e$.

Now we formulate the Theorem of isomorhism for real factors of type III_0.

Theorem 2.2.14. *Let $(Q_1, \theta_1), (Q_2, \theta_2)$ be a couple satisfying the conditions of Theorem 2.2.7, let λ_1 and λ_2 be numbers such that $\tau_j \cdot \theta_j \leq \lambda_j \tau_j$, where τ_1 and τ_2 are faithful normal semifinite traces on Q_1 and Q_2 respectively. Let $\Re_1 = Q_1 \times_{\theta_1} Z$ and $\Re_2 = Q_2 \times_{\theta_2} Z$ be real factors of type III_0. Then $\Re_1$ is isomorphic to $\Re_2$ if and only if there exist nonzero projections $e_1 \in Z(Q_1)$ and $e_2 \in Z(Q_2)$ and a *-isomorphism $J : e_1 Q_1 \to e_2 Q_2$ such that*

$$J \cdot \theta_{1,e_1} \cdot J^{-1} \cdot \theta_{2,e_2}^{-1} \in Int(e_2 Q_2).$$

Proof. Let (Q, θ) be a couple satisfying the conditions of Theorem 2.2.7, put $\Re = Q \times_\theta Z$, and let τ be a faithful normal semifinite traces on Q such that $\tau \cdot \theta \leq \lambda_0 \tau$, where $0 < \lambda_0 < 1$. Let $I : Q \to I(Q) \subset \Re$ be the canonical imbedding of Q into $\Re$, let $E : \Re \to I(Q)$ be the corresponding conditional expectation, $u = \lambda_\theta(1) \in \Re$.
First we prove a few Lemmas.

Lemma 2.2.15. *Let $\varphi = \tau \cdot I^{-1} \cdot E$ be a faithful normal semifinite weight on $\Re_+$, $\overline{\varphi}$ an α-invariant extension of φ on $\mathcal{U}_+$, and let $\sigma = \sigma^\varphi$ be an action of modular *-automorphism group σ^φ on $\mathcal{U}$ associated with φ. Then*

(1) $\overline{\varphi}$ is lacunary weight of infinite multiplicity;

(2) for all $x \in \Re$, we have

$$Sp_\sigma(x - E(x)) \subset (\lambda_0, \frac{1}{\lambda_0}), \quad and \quad \Re^\sigma = \mathcal{U}_{\overline{\varphi}} \cap \Re = I(Q);$$

(3) $Sp_\sigma(u) \subset (1, +\infty)$;

(4) if v is a unitary from $\Re$ such that

$$v \Re^\sigma v^* = \Re^\sigma, \; v^* \Re^\sigma v = \Re^\sigma,$$

$\Re$ is generated by $\Re^\sigma$ and u, and $Sp_\sigma(v) \subset (1, +\infty)$, then $u^*v \in \Re^\sigma$.

Proof. Let $\overline{\tau}$ be the α-invariant trace on $\mathcal{U}(Q)$ obtained by extending of the trace τ, let $\overline{E}$ be the conditional expectation from $\mathcal{U}$ on $\mathcal{U}(I(Q))$ which is the linear extension of E and let $\overline{I}$ be an isomorphism from $\mathcal{U}(Q)$ into $\mathcal{U}(I(Q))$ obtained as the linear extension of I. Then for $\mathcal{U}(Q)$, $\overline{\tau}$, $\overline{\theta}$, and λ_0 the conditions of Lemma 5.4.3 from [C 1] are fulfilled. From this lemma the statements (1) - (3) follow directly.

Proof of the statement (4). By Lemma 5.4.3 from [C 1], $u^*v \in \mathcal{U}_{\overline{\varphi}}$. Since $u, v \in \Re$, we have $u^*v \in \Re^\sigma$. ■

Lemma 2.2.16. *Let $e \neq \mathbf{I}$ be a nonzero projection from $Z(Q)$, let $\{e_n\}_{n=1}^\infty \subset Z(Q)$ be the sequence of projections corresponding to e, θ_e - the corresponding *-automorphism of eQ, let $\Re_0 = eQ \times_{\theta_e} Z$ and $u_0 = \lambda_{\theta_e}(1) \in \Re_0$. Further let I_0 be the canonical imbedding of eQ into $\Re_0$, E_0 - the conditional expectation from $\Re_0$ into $I(eQ)$, σ_0 the restriction of action of $\sigma^{\overline{\varphi_0}}$ onto eQ. Suppose that $\varphi_0 = \tau|_{eQ} \cdot I^{-1} \cdot E_0$ is a faithful normal semifinite weight on $(\Re_0)_+$, $\overline{\varphi_0}$ - the extension of φ_0 on $\mathcal{U}(eQ)$. Then there exists a *-isomorphism J_0: $\Re_0 \to I(e)\Re I(e)$, such that*

(1) $J_0 \cdot I_0(x) = I(x)$, *for all* $x \in eQ$;

(2) $J_0(E_0(x)) = I(J_0(x))$, *for all* $x \in \Re_0$;

(3) $J_0(u_0) = \sum_{n=1}^\infty u^n(e_n)$;

(4) $\varphi(J_0(x)) = \varphi_0(x)$, *for all* $x \in (\Re_0)_+$;

(5) $J_0(\sigma_{0,t}(x)) = \sigma_t(J_0(x))$, *for all* $x \in \Re_0$, *for all* $t \in \mathbf{R}$.

Proof. Put

$$f = I(e), \quad u_1 = \sum_{n=1}^\infty u^n I(e_n).$$

By the construction the element u_1 is a unitary, and

$$u_1 I(x) u_1^* = I(\theta_e(x)), \quad \text{for all} \quad x \in eQ.$$

Let $\Re_1$ be the smallest real W^*-subalgebra of $f\Re f$, containing the set $I(eQ) = I(Q) \cap (fQf)$ and the element u. By Lemma 5.4.4 from [C 1] we see that the smallest W^*-algebra containing u_1 and

$$\mathcal{U}(I(eQ)) = I(eQ) + iI(eQ)$$

coincides with the W^*-algebra $f\mathcal{U}f = f\Re f + if\Re f$. Hence, $\Re_1 = f\Re f$. Then by Lemma 4.1.2 from [C 1] and by Proposition 2.1.1 (see also the proof of Theorem 2.2.8) there exists a *-isomorphism $J_0 : \Re_0 \to f\Re f$ such that the statements (1) - (3) are valid for J_0.

Let the elements $x \in (\Re_0)_+$, $y \in (eQ)_+$ be such that $E_0(x) = I_0(y)$. By the definition of φ_0, we have

$$\varphi_0(x) = \tau(y), \quad \text{and} \quad \varphi(J_0(x)) = \tau_0 I^{-1}(E(J_0(x))).$$

Hence,

$$E(J_0(x)) = J_0(I_0(y)), \qquad I^{-1}(J_0(I_0(y))) = y,$$

and thus $\varphi(J_0(x)) = \tau(y)$. This proves the statement (4). The statement (5) follows from the fact that $\varphi_0 \cdot J_0 = \varphi_f$ (where φ_f is the restriction of φ_f on $f\Re f$), and from Lemma 3.2.6 from [C 1]. ∎

Lemma 2.2.17. *Let φ be a faithful normal strictly semifinite α-invariant lacunary weight of infinite multiplicity on $\mathcal{U}_+$, $\sigma = \sigma^\varphi$, and suppose that a_1, a_2 are nonzero projections from $Z(\mathcal{U}_\varphi) \cap \Re$ such that $\{0\}$ is an isolated point in each of $Sp\sigma^{a_1}$ and $Sp\sigma^{a_2}$; let $\varepsilon > 0$ be a fixed positive number. Then there exists a partial isometry $v \in \Re$ such that the projections $b_1 = v^*v, b_2 = vv^*$ are contained in $Z(\mathcal{U}_\varphi) \cap \Re$, $b_i \leq a_i$ and*

$$Sp_\sigma(v) - Sp_\sigma(v) \subset [-\varepsilon, \varepsilon].$$

Proof. By Lemmas 2.2.6 and 2.2.9 we may suppose without loss of generality that

$$Sp\sigma^{a_i} \cap [-\varepsilon, \varepsilon] = \{0\}, \quad i = 1, 2.$$

In the proof of Lemma 5.4.5 from [C 1] the element $y \in \mathcal{U}$ was costructed such that $a_2 y = y = y a_1$ and

$$Sp_\sigma(y) = Sp_\sigma(y) \subset [-\varepsilon, \varepsilon].$$

The element y has a decomposition: $y = y_1 + i y_2$, where $y_1, y_2 \in \mathfrak{R}$. Put $e = s_r(y_1), e' = s_l(y_1)$. For y_1, by corollary 2.2.2, we have: $Sp_\sigma(y_1) \subseteq Sp_\sigma(y)$ and

$$Sp_\sigma(y_1) - Sp_\sigma(y_1) \subseteq Sp_\sigma(y) - Sp_\sigma(y) \subset [-\varepsilon, \varepsilon].$$

Then y_1 satisfies the conditions of Lemma 2.2.9 and thus there exists a partial isometry $v \in \mathfrak{R}$ such that $b_1 = v^* v$ is the relative central support of e in $\mathcal{U}_\varphi$, $b_2 = v v^*$ is the relative central support of e' in $\mathcal{U}\varphi$, and

$$Sp_\sigma(v) - Sp_\sigma(v) \subset [-\varepsilon, \varepsilon].$$

It is easy to see that $b_1 \leq a_1$, as $e \leq a_1$, $a_1 \in Z(\mathcal{U}_\varphi)$ and that $b_2 \leq a_2$, as $e' \leq a_2$ and $a_2 \in Z(\mathcal{U}_\varphi)$. ∎

Lemma 2.2.18. *Let $\mathfrak{R}$ be a real factor of type III_0, φ_1 and φ_2 - faithful normal strictly semifinite weights on $\mathfrak{R}_+$ such that their α-invariant extensions $\overline{\varphi}_1$ and $\overline{\varphi}_1$ are lacunary weights of infinite multiplicity on $\mathcal{U}_+$, let $\varepsilon > 0$ be a fixed number. Put*

$$\mathfrak{R}_{\varphi_i} = \mathcal{U}_{\overline{\varphi}_i} \cap \mathfrak{R}, \quad i = 1, 2.$$

*Then there exist projections $e_1 \in Z(\mathfrak{R}_{\varphi_1})$, $e_2 \in Z(\mathfrak{R}_{\varphi_2})$ and a partial isometry $v \in \mathfrak{R}$, such that $v^* v = e_1, v v^* = e_2$, and the *-isomorphism $j(\cdot) = v \cdot v^*$ from $\mathfrak{R}_{\varphi_1}$ onto $\mathfrak{R}_{\varphi_2}$; maps $\mathfrak{R}_{\varphi_1} \cap (e_1 \mathfrak{R} e_1)$ onto $\mathfrak{R}_{\varphi_2} \cap (e_2 \mathfrak{R} e_2)$, and moreover*

$$\ln(Sp_{\sigma_2}(j(x))) \subset \ln(Sp_{\sigma_1}(x)) + [-\varepsilon, \varepsilon],$$

$$\ln(Sp_{\sigma_1}(x)) \subset \ln(Sp_{\sigma_2}(j(x))) + [-\varepsilon, \varepsilon],$$

for all $x \in e_1 \mathfrak{R} e_1$, where σ_i is the restriction of $\sigma^{\overline{\varphi}_i}$ on $e_i \mathfrak{R} e_i$, $i = 1, 2$.

The proof of the Lemma follows the scheme of the proof of Lemma 5.4.6 from [C 1], if we apply Lemma 2.2.17.

Lemma 2.2.19. *Let* Q *be a* σ*-finite real* W^**-algebra,* $\theta_1, \theta_2 \in Aut(Q)$. *If* $\theta_1\theta_2^{-1} \in Int(Q)$ *then* $Q \times_{\theta_1} Z$ *is *-isomorphic to* $Q \times_{\theta_2} Z$.

Proof. Let I_i be the canonical imbedding of Q into $Q \times_{\theta_i} Z$, $u_i = \lambda_{\theta_i}(1)$ $i = 1, 2$. Let u be a unitary from Q such that $\theta_2(x) = u\theta_1(x)u^*$, for every $x \in Q$. Put $u_2' = I_1(u)u_1$. Then $Q \times_\theta Z$ is generated by $I_1(Q)$ and u_2', and for every $x \in Q$ we have:

$$I_1(\theta_2(x)) = u_2{}' I_1(x) u_2{}'^*.$$

Then following step by step to the proof of Lemma 4.1.2 from [C 1] we obtain that $Q \times_{\theta_1} Z$ and $Q \times_{\theta_2} Z$ are *-isomorphic. ∎

Now we shall prove the Theorem 2.2.14.

If

$$J \cdot \theta_{1,e_1} \cdot J^{-1} \cdot \theta_{2,e_2}^{-1} \in Int(e_2Q_2),$$

then by Lemma 2.2.19 $e_1Q \times_{\theta_{1,e_1}} Z$ and $e_2Q \times_{\theta_{2,e_2}}$ are *-isomorphic. By Lemma 2.2.16 the real W^*-algebra $e_iQ_i \times_{\theta_{i,e_i}} Z$ is *-isomorphic to the real W^*-algebra $e_i(e_iQ_i \times_{\theta_{i,e_i}} Z)e_i$, $i = 1, 2$. Since $Q_i \times_{\theta_i} Z$ are real factors of type III, we have $e_i \sim \mathbf{1}$ in $Q_i \times_{\theta_i} Z$, hence it follows that the real W^*-algebras $e_i(Q_i \times_{\theta_i} Z)e_i$ and $Q_i \times_{\theta_i} Z$ are *-isomorphic, $i = 1, 2$. Therefore $Q_1 \times_{\theta_1} Z$ is *-isomorphic to $Q_2 \times_{\theta_2} Z$.

Now we shall prove the converse. We may suppose without loss of generality that $\Re_1 = \Re_2 = \Re$. Let φ_1, and φ_2 be faithful normal strictly semifinite weights on $\Re_+$ such that the weights $\overline{\varphi}_1$ and $\overline{\varphi}_2$ (α-invariant extensions of φ_1 and φ_2 respectively) are lacunary weights of infinite multiplicity. We choose the number $\varepsilon > 0$ from in Lemmas 2.2.16 - 2.2.18 such that $|\ln \lambda_i| > \varepsilon$, where λ_i are such that $\tau_i \cdot \theta_i \leq \lambda_i \tau_i$, $i = 1, 2$. We shall denote the real W^*-algebra $\mathcal{U}_{\overline{\varphi}_i} \cap \Re$ by $\Re_{\varphi_i}$, $i = 1, 2$. By Lemma 2.2.18 there exist $f_i \in Z(\Re_{\varphi_1})$ and $f_2 \in Z(\Re_{\varphi_2})$, and a *-isomorphism $j : \Re_{\varphi_1} \to \Re_{\varphi_2}$ such that

$$j(\Re_{\varphi_1} \cap (f_1 \Re f_1)) = \Re_{\varphi_2} \cap (f_2 \Re f_2)$$

and

$$\ln(Sp_{\sigma_2}(j(x))) \subset \ln(Sp_{\sigma_1}(x)) + [-\varepsilon, \varepsilon].$$

By Lemma 2.2.15 (2), $\Re_{\varphi_k} = I^{(k)}(Q_k)$, where $I^{(k)}$ is the canonical imbedding of Q_k into $\Re$; thus there exist projections $e_k \in Z(Q_k)$, such that $I^{(k)}(e_k) = f_k$, $k = 1, 2$. Therefore

$$I^{(1)}(e_1 Q_1) = \Re_{\varphi_1} \cap (f_1 \Re f_1), \quad I^{(2)}(e_2 Q_2) = \Re_{\varphi_2} \cap (f_2 \Re f_2),$$

and there exists a *-isomorphism $J :$ $J : e_1 Q_1 \to e_1 Q_2$ such that

$$I^{(2)}(J(x)) = j(I^{(1)}(x))$$

for all $x \in e_1 Q_1$. By Lemma 2.2.16, there exists a unitary $v \in f_1 \Re f_1$ such that

(1) $I^{(1)}(Q_{1,e_1}(x)) = vI^{(1)}(x)v^*$, for all $x \in e_1 Q_1$;

(2) $f_1 \Re f_1$ is the smallest real W^*-algebra containing $I^{(1)}(e_1 Q_1)$ and v;

(3) $Sp_{\sigma_1}(v) \subset (\frac{1}{\lambda_1}, +\infty)$.

Then $j(v) \in f_2 \Re f_2$ and $Sp_{\sigma_2}(j(v)) \subset (1, +\infty)$, since

$$\ln(Sp_{\sigma_1}(v)) \subset [-\ln \lambda_1, +\infty),$$

and therefore

$$\ln(Sp_{\sigma_2}(j(v))) \subset (0, +\infty).$$

Moreover,

$$j(v)(f_2(\Re_{\varphi_2})f_2)j(v)^* = f_2(\Re_{\varphi_2})f_2.$$

Lemma 2.2.15 (4) shows that there exists a unitary $a \in f_2(\Re_{\varphi_2})f_2$, such that $u = aj(v)$ satisfies the condition:

$$I^{(2)}(\theta_{2,e_2}(x)) = uI^{(2)}(x)u^*,$$

for all $x \in e_2 Q_2$. Since $I^{(2)}(x) = j(I^{(1)}(J^{-1}(x)))$ for all $x \in e_2 Q_2$, we have

$$I^{(2)}(\theta_{2,e_2}(x)) = j(I^{(1)}(J^{-1}(\theta_{2,e_2}(x)))) =$$

$$aj(vI^{(1)}(J^{-1}(x))v)a^* = aj(I^{(1)}\theta_{1,e_1}(J^{-1}(x))))a^*,$$

hence it follow that the *-automorphism $\theta_{1,e_1} \cdot J^{-1} \cdot \theta_{2,e_2}^{-1} \cdot J$ is inner on $e_1 Q_1$. ■

2.3. Periodic antiautomphisms and automorphisms of complex and real type II_1 factors

From results of Sections 2.1 and 2.2 we see that the classification of real type III factors is tightly connected with the classification of *-automorphisms and *-antiautomorphisms of real and complex W^*-algebras of type II_∞. Further, our goal is the description of outer conjugacy classes of *-automorphisms and *-antiautomorphisms of injective factors.

First we shall consider periodic *-automorphisms and periodic *-antiautomorphisms of injective real and complex II_1 factors.

Let M be a complex or a real W^*-algebra, and let α be a *-automorphism or *-antiautomorphism of M. We remind that, two *-automorphisms or *-antiautomorphisms α and β are said to be *conjugate* (or *outer conjugate*), if $\alpha = \theta \cdot \beta \cdot \theta^{-1}$ (respectively, $Adu \cdot \alpha = \theta \cdot \beta \cdot \theta^{-1}$) for some *-automorphism θ (and an inner *-automorphism $Adu(x) = uxu^*$). A *-automorphism (or a *-antiautomorphism) α is said to be *periodic* with period p if $\alpha^p(x) = x$ for all $x \in M$, where p is a positive integer. It is easy to see that in this case the smallest positive integer $p_0 = p_0(\alpha)$ and a unitary element u in M exist such that $\alpha^{p_0} = Adu$; besides, if M is a factor, then we have the relation $\alpha(u) = \gamma(\alpha)u^*$, $\gamma(\alpha) \in \mathbf{C}$, $|\gamma(\alpha)| = 1$ provided that α is a *-antiautomorphism. However, for a real factor M, if α is a *-automorphism, we have the relation $\alpha(u) = \gamma(\alpha)u$, $\gamma(\alpha) = \pm 1$, and if α is a *-antiautomorphism, we have $\alpha(u) = \gamma(\alpha)u^*$, $\gamma(\alpha) = \pm 1$. A periodic *-automorphism or *-antiautomorphism α is said to be *purely periodic* if $p = p_0$, where p is the period of α.

Let M be a factor, and let α be a periodic *-automorphism of M. It is easy to see that $\gamma(\alpha)^{p_0} = 1$. Since $\gamma(\alpha)$ is the p_0 - th root of 1, denote by q the order of complex number $\gamma(\alpha)$. Put $p_m(\alpha) = q \cdot p_0$. The number $p_m(\alpha)$ is called the *minimal period* of the mapping α. If $p = p_m(\alpha)$, then the *-automorphism α is said to be *minimally periodic* (see [C 5]). Let $\Re$ be a real factor, α - a periodic *-automorphism of $\Re$ with period p. Put

$$p_m(\alpha) = \begin{cases} p_0(\alpha), & \text{if } \gamma(\alpha) = 1 \\ 2p_0(\alpha), & \text{if } \gamma(\alpha) = -1 \end{cases}$$

If $p = p_m(\alpha)$, then the *-automorphism α of $\Re$ is said to be *minimally periodic.*

Proposition 2.3.1.

*(i) Let M be a continuous factor, α - a periodic *-antiautomorphism of M with period p. The *-antiautomorphism α is purely periodic if and only if*

$$M_s^\alpha = \{x \in M : \alpha(x) = x = x^*\}$$

is a continuous JW-factor not isomorphic to the self-adjoint part of a W^*-algebra.

*(ii) Let $\Re$ be a continuous real factor, α a periodic *-automorphism of $\Re$ with period p. If the *-automorphism α is minimally periodic, then*

$$\Re^\alpha = \{x \in \Re : \alpha(x) = x^*\}$$

is a real W^*-subfactor.

Proof. (i). Let α be purely periodic and put $m = p/2$. Then α^2 is a minimally periodic *-automorphism with period m. By theorem 2.5 (c) of [C 5],

$$M^{\alpha^2} = \{x \in M : \alpha^2(x) = x\}$$

is a factor. Then it is clear that $M^\alpha = M^{\alpha_1}$, where $\alpha_1 = \alpha|_{M^{\alpha^2}}$. Since

$$M^{\alpha_1} = \{x \in M : \alpha_1(x) = x\} = \{x \in M^{\alpha^2} : \alpha_1(x) = x\},$$

we obtain $M_s^{\alpha_1} = M_s^\alpha$. By Theorem 1.5.2 it follows that $M_s^{\alpha_1} = M_s^\alpha$ is a continuous JW-factor not isomorphic to the self-adjoint part of a W^*-algebra, because α_1 is an involutive *-antiautomorphism.

Conversely, let $N = M^{\alpha^2}$ be a W^*-algebra. It is easy to see that $A = \{x \in N_s : \alpha(x) = x\}$ is a JW-algebra, and its enveloping W^*-algebra $\mathcal{U}(A)$ coincides with N. By the assumption of proposition the JW-algebra A isn't isomorphic to self-adjoint the part of a W^*-algebra. Therefore, again by Theorem 1.5.2, we obtain that $\mathcal{U}(A)$ is a W^*-factor. From Theorem 2.5 (a) of [C 5] it follows that the *-automorphism α^2 is minimally periodic and therefore α is a purely periodic *-antiautomorphism.

(ii). Let $\overline{\alpha}$ be the *-automorphism of $\mathcal{U}(\mathfrak{R}) = \mathfrak{R} + i\mathfrak{R}$, which is the linear extension of α : $\overline{\alpha}(ix) = i\alpha(x)$, $x \in \mathfrak{R}$. Using Theorem 1.6 of [S 5] we obtain that $p_0(\overline{\alpha}) = p_0(\alpha)$. Moreover,it follows directly that $\gamma(\overline{\alpha}) = \gamma(\alpha)$. Hence $p_m(\overline{\alpha}) = p_m(\alpha)$. Therefore $\overline{\alpha}$ is minimally periodic, i.e. $\mathcal{U}(\mathfrak{R})^{\overline{\alpha}}$ is a factor. Since $\mathcal{U}(\mathfrak{R})^{\overline{\alpha}} = \mathfrak{R}^\alpha + i\mathfrak{R}^\alpha$, it follows that $\mathfrak{R}^\alpha$ is a real W^*-subfactor. ■

Let M be a continuous factor, $\mathfrak{R}$ - a continuous real factor. Using Proposition 2.3.1, similarly to the proofs of corollary 2.6 (a) in [C 5] and Lemma 1.6 in [G 3], we obtain the following statements.

Proposition 2.3.2. *Let α be a purely (respectively minimal) periodic *-antiautomorphism of M (respectively*-automorphism of $\mathfrak{R}$) with period p. Then*

(i) A unitary $u \in M$ is of the form $u = v\alpha(v)$, for an apprepriate unitary v in M, if and only if

$$u\alpha(u^*)\alpha^2(u)...\alpha^{p-1}(u^*) = \mathbf{I};$$

(ii) A unitary $u \in \mathfrak{R}$ is of the form $u = v^*\alpha(v)$, for an appropriate unitary v in M, if and only if

$$u\alpha(u)\alpha^2(u)...\alpha^{p-1}(u) = \mathbf{I},$$

*in this case the *-antiautomorphisms (respectively *-automorphisms) α and $Adu \cdot \alpha$ are conjugate.*

Proposition 2.3.3.

(i) Let $\alpha \in Ant(M)$, and let $\alpha^p = Adu$, $\alpha(u) = u^*$. Then there exists a unitary $v \in M$ such that $u^* = v\alpha(v^*)\alpha^2(v)...\alpha^{p-1}(v^*)$;

(ii) Let $\alpha \in Aut(\Re)$, and let $\alpha^p = Adu$, $\alpha(u) = u$. Then there exists a unitary $v \in \Re$ such that $u^* = v\alpha(v)\alpha^2(v)...\alpha^{p-1}(v)$;

(iii) $\|v - \mathbf{1}\|_\psi \leq 2\|u - \mathbf{1}\|$, for each $\psi \in M_*^+$ (respectively $\psi \in \Re_*^+$), where $\|x\|_\psi = \psi(x^*x)^{1/2}$.

Let ω be a free ultrafilter on N (i.e. a maximal filter which doesn't contain finite sets). A sequence $\{x_n\}_{n\in \mathrm{N}}$ of elements in M is called *central* (respectively *ω-central*), if it is the element of the C^*-algebra $L^\infty(\mathrm{N}, M)$, and for each $\psi \in M_*$ we have

$$\|[\psi, x]\| \rightarrow 0, \quad \text{when } n \rightarrow \infty \quad (\text{respectively } n \rightarrow \omega),$$

where $[\psi, x](y) = \psi(xy) - \psi(yx)$. Let $\oplus_\infty M$ be the direct sum of a countable number of copies of M and let

$$J_\omega = \{(x_n) \in \oplus_\infty M : x_n \rightarrow 0 \ * -\text{strongly}, \text{ when } n \rightarrow \omega\},$$
$$\tilde{M} = \{(x_n) \in \oplus_\infty M : x_n = x, \text{ for all } n\}.$$

Let ρ be the canonical homomorphism of $\oplus_\infty M$ onto $\oplus_\infty M/J_\omega$. Put

$$M_\omega = (\oplus_\infty M/J_\omega) \bigcap \rho(\tilde{M})'.$$

It is known that M_ω is the algebra of all equivalence classes of ω-centralizing sequences in M (see [D 1, p. 31]). Similarly we define $\Re_\omega$, where $\Re$ is a real W^*-algebra, moreover, in [S 5] it is proved that the *-algebra of central sequences $\Re_\omega$ is a real W^*-algebra and

$$\mathcal{U}(\Re)_\omega = \Re_\omega + i\Re_\omega.$$

Let α be a *-automorphism (or *-antiautomorphism) of M (respectively of $\Re$). It is easy to see that the mapping: $(x_n)_{n\in\mathrm{N}} \rightarrow (\alpha(x_n))_{n\in\mathrm{N}}$ defines a *-automorphism (or *-antiautomorphism) α_ω of M_ω (respectively of $\Re_\omega$).

Let $\mathfrak{R}$ be a real W^*-algebra, $\mathcal{K}$ a real subfactor in $\mathfrak{R}$; and put $\mathcal{K}'_{\mathfrak{R}} = \mathcal{K}' \cap \mathfrak{R}$. Suppose that $\mathcal{R}(\mathcal{K}, \mathcal{K}'_{\mathfrak{R}})$ is the smallest real factor containing $\mathcal{K}$ and $\mathcal{K}'_{\mathfrak{R}}$. It is easy to see that $\mathcal{R}(\mathcal{K}, \mathcal{K}'_{\mathfrak{R}})$ coincides with $\mathfrak{R}$. Let φ be a faithful normal state on $\mathfrak{R}$, let σ_t be the modular automorphism group on $\mathcal{U}(\mathfrak{R})$, associated with $\overline{\varphi}$, where $\overline{\varphi}$ is the α-invariant linear extension the state φ on $\mathcal{U}(\mathfrak{R})$.

Lemma 2.3.4. *If the subfactor $\mathcal{K}$ is invariant under the modular automorphism group σ_t, then the map*

$$\pi : \quad \sum_{j=1}^{n} x_j \otimes y_j \in \mathcal{K} \otimes \mathcal{K}'_{\mathfrak{R}} \longrightarrow \sum_{j=1}^{n} x_j y_j \in \mathcal{R}(\mathcal{K}, \mathcal{K}'_{\mathfrak{R}})$$

can be extended to an isomorphism of $\mathcal{K} \otimes \mathcal{K}'_{\mathfrak{R}}$ onto the W^*-algebra $\mathcal{R}(\mathcal{K}, \mathcal{K}'_{\mathfrak{R}})$.

Proof. Let $\mathcal{R}_0(\mathcal{K}, \mathcal{K}'_{\mathfrak{R}})$ denote the algebra generated algebraically by $\mathcal{K}$ and $\mathcal{K}'_{\mathfrak{R}}$. It is known (see for example [Dix 1, Exercise 6, p. 29]) that the map

$$\pi_1 : \quad \sum_{j=1}^{n} x_j \otimes y_j \in \mathcal{K} \otimes \mathcal{K}'_{\mathfrak{R}} \longrightarrow \sum_{j=1}^{n} x_j y_j \in \mathcal{R}_0(\mathcal{K}, \mathcal{K}'_{\mathfrak{R}})$$

is an isomorphism. Since $\mathcal{K}$ is invariant under the modular automorphism group σ_t, then by corollary 1 from [T 3] we have $\varphi(xy) = \varphi(x)\varphi(y)$ for every $x \in \mathcal{K}$, $y \in \mathcal{K}'_{\mathfrak{R}}$. It is not difficult to conclude from this that the map π_1 is extended to an isomorphism of $\mathcal{K} \otimes \mathcal{K}'_{\mathfrak{R}}$ onto $\mathcal{R}(\mathcal{K}, \mathcal{K}'_{\mathfrak{R}})$. ∎

Lemma 2.3.5. *Let $\mathcal{K}$ be a real type I_n subfactor of $\mathfrak{R}$ with matrix units $\{e_{i,j},\ i,j = \overline{1,n}\}$ and let $\tau_{\mathcal{K}}$ be the normalized trace on $\mathcal{K}$. Then for any $\psi \in \mathfrak{R}_*$ one has*

$$\|\psi \ - \ \psi|_{\mathcal{K}'_{\mathfrak{R}}} \otimes \tau_{\mathcal{K}}\| \ \leq \ n^2 \sup_{i,j} \|[e_{ij}, \psi]\|.$$

Proof of the Lemma is carried out easily, similarly to the proof of Lemma 2.3.5 in [C 2].

Lemma 2.3.6. *Let* $\{n_m\}_{m\in\mathrm{N}}$ *be a sequence of positive integers* ≥ 2, $\{\mathcal{K}_m\}_{m\in\mathrm{N}}$ *a sequence of pairwise commuting subfactors of* $\mathfrak{R}$ *with* $\mathcal{K}_m$ *of type* I_{n_m} *for all* $m \in \mathrm{N}$. *Let* $\mathcal{K}$ *be a real subfactor generated by* $\{\mathcal{K}_m, m \in \mathrm{N}\}$, *and* $\{\psi_j\}_{j\in\mathrm{N}}$ *a countable total subset of* $\mathfrak{R}_*$. *Assume that for all* $j \in \mathrm{N}$ *one has*

$$\sum_m \|\psi_j - \psi_j|_{(\mathcal{K}_m)'_{\mathfrak{R}}} \otimes \tau_{\mathcal{K}_m}\| < \infty.$$

Then $\mathcal{K}$ *is a real subfactor of type* II_1 *of* $\mathfrak{R}$, *and* $\mathfrak{R}$ *is equal to the tensor product of* $\mathcal{K}$ *and its relative commutant* $\mathcal{K}'_{\mathfrak{R}}$.

Proof. Let μ_m be the Haar measure on the unitary group of $\mathcal{K}_m$, $m \in \mathrm{N}$, such that $\mu(\mathbf{I}) = 1$. For $x \in \mathfrak{R}$ define

$$E_m(x) = \int uxu^* \, d\mu_m(u),$$

where u is a unitary in $\mathfrak{R}$. It is easy to see that the map $E_m : \mathfrak{R} \longrightarrow (\mathcal{K}_m)'_{\mathfrak{R}}$ is a faithful normal conditional expectation of $\mathfrak{R}$ on the relative commutant $(\mathcal{K}_m)'_{\mathfrak{R}}$ of $\mathcal{K}_m$, and if identifying $\mathfrak{R}$ with $\mathcal{K} \otimes \mathcal{K}'_{\mathfrak{R}}$, it coincides with $\tau_{\mathcal{K}_m} \otimes id$.

Consider the transposition E_m^* of E_m in $\mathfrak{R}_*$ which is a projection of norm 1, and for each $\psi \in \mathfrak{R}_*$ we have

$$\psi \cdot E_m = \psi|_{(\mathcal{K}_m)'_{\mathfrak{R}}} \otimes \tau_{\mathcal{K}_m}.$$

So we can rewrite the hypothesis of the Lemma as

$$\sum_m \|E_m^* \cdot \psi_j - \psi_j\| < \infty, \qquad j \in \mathrm{N}.$$

Now the maps E_m , $m \in \mathrm{N}$ obviously commute pairwise because Adu and Adv commute for each unitary u in $\mathcal{K}_m$, and unitary v in $\mathcal{K}_n$, $m \neq n$. Hence, $\{E_m\}_{m\in\mathrm{N}}$ commute pairwise, and therefore, $\{E_m^*\}_{m\in\mathrm{N}}$ also commute pairwise. Since for any $j \in \mathrm{N}$ the sequence

$$\left\{\psi_j^n = \psi_j\left(\prod_1^n E_m^*\right)\right\}_{n\in\mathrm{N}}$$

satisfies the condition

$$\sum_{n=1}^{\infty} \|\psi_j^{n+1} - \psi_j^n\| \leq \sum_{n=1}^{\infty} \|E_{n+1}^* \psi_j - \psi_j\| < \infty,$$

thus the sequence $\{\prod_1^n E_m^*\}_{n\in\mathrm{N}}$ converges pointwise in norm, i.e.

$$\prod_1^{\infty} E_m^* = E^*.$$

It follows that the product $\{\prod_1^n E_m\}_{n\in\mathrm{N}}$ converges pointwise weakly to the transpose E. By the construction we have

$$(\prod_1^n E_j)(x) \in \mathcal{K}_m, \; x \in \mathfrak{R}, \; m < n;$$

$$E_m y = y, \; y \in \mathcal{K}'_{\mathfrak{R}} = \bigcap_m (\mathcal{K}_m)'_{\mathfrak{R}}, \; m \in \mathrm{N}.$$

Hence, $Ey = y$. Thus E is a weakly continuos projection of norm 1 and $E : \mathfrak{R} \longrightarrow \mathcal{K}'_{\mathfrak{R}}$. If $x \in \mathfrak{R}$ and u is a unitary in $\mathcal{K}$, then by the construction we have $E(uxu^*) = E(x)$.

Now for any faithful normal state φ on $\mathcal{K}'_{\mathfrak{R}}$, we have that $\psi = \varphi \cdot E$ is a normal state on $\mathfrak{R}$ such that $\psi(uxu^*) = \psi(x), \; x \in \mathfrak{R}$. We shall prove that the map E is faithful. Indeed, let $e = s(\psi)$ be the support of ψ. Then by the definition of φ and ψ we have $e \in \mathcal{K}'_{\mathfrak{R}}$. Since $E(e) = e$, we obtain $\psi(\mathbf{1} - e) = \psi(\mathbf{1} - e) = 0$ and $e = \mathbf{1}$. We have shown that E is faithful and that $\mathcal{K}$ is a finite real factor, because $(\varphi \cdot E)|_{\mathcal{K}}$ is a faithful normal trace on $\mathcal{K}$ (see [Dix 1, p. 271]).

Now let $\overline{\psi}$ be a state on $\mathcal{U}(\mathfrak{R})$ with $\overline{\psi}|_{\mathfrak{R}} = \psi$ and let σ_t be the modular automorphism group of $\mathcal{U}(\mathfrak{R})$, associated with $\overline{\psi}$. Then by Lemma 2.3.4 we have

$$\mathcal{K} \otimes \mathcal{K}'_{\mathfrak{R}} = \mathfrak{R}. \quad \blacksquare$$

Definition 2.3.7. A real W^*-algebra $\mathfrak{R}$ in $B(H)$ is called *injective*, if there exists a projection P from $B(H)$ on $\mathfrak{R}$ such that $\|P\| = 1, P(\mathbf{1}) = \mathbf{1}$;

Proposition 2.3.8. *A real W^*-algebra $\mathfrak{R}$ is injective if and only if the W^*-algebra $\mathcal{U}(\mathfrak{R})$ is injective.*

Proof. Let $\mathfrak{R}$ be injective, i.e. there exist the projection P from $B(H)$ on $\mathfrak{R}$ with $\|P\| = 1$, and $P(\mathbf{I}) = \mathbf{I}$. Then by Theorem 3.2 from [HS] there exist a projection $\overline{P}$ from $B(H)$ on $\mathcal{U}$ with $\|\overline{P}\| = 1$, and $\overline{P}(\mathbf{I}) = \mathbf{I}$, i.e., $\mathcal{U}$ is injective.

Conversely, let $\mathcal{U}$ be injective and let $\overline{P}$ be a projection from $B(H)$ on $\mathcal{U}$ with $\|\overline{P}\| = 1$, and $\overline{P}(\mathbf{I}) = \mathbf{I}$. Consider the map

$$E(x) = \frac{1}{2}(x + \alpha(x^*)), \quad x \in \mathcal{U}.$$

It is easy to see that E is a projection from $\mathcal{U}$ on $\mathfrak{R}$. Indeed, let $y = E(x)$, then we have

$$\alpha(y) = \frac{1}{2}\alpha(x + \alpha(x^*)) = \frac{1}{2}(\alpha(x) + \alpha^2(x^*)) = \frac{1}{2}(\alpha(x) + x^*) = y^*,$$

i.e. $y \in \mathfrak{R}$. It is clear that $\|E\| = 1$, $E(\mathbf{I}) = \mathbf{I}$. Then the map $P = E \cdot \overline{P}$ is a projection from $B(H)$ on $\mathfrak{R}$, moreover, $\|P\| = 1$, and $P(\mathbf{I}) = \mathbf{I}$, i.e. $\mathfrak{R}$ is injective. ∎

Let M be an injective factor, $\mathfrak{R}$ - a real injective factor and let α be a purely periodic $*$-antiautomorphism of M (respectively a minimally periodic $*$-automorphism $\mathfrak{R}$) with period p. Let ψ be a normal faithful state, $\psi_1, ..., \psi_q \in M_*$, (respectively $\in \mathfrak{R}_*$) and $\varepsilon > 0$.

Lemma 2.3.9. *There exist a unitary u in M (respectively, $u \in \mathfrak{R}$) and a system of matrix units $\{(e_{ij}), i,j = 1, ..., p\}$ of M (respectively, $\mathfrak{R}$) such that*

a) $\|[\psi_r, e_{ij}]\| \leq \varepsilon, \quad r = 1, ..., q; \quad i,j = 1, ..., p;$

b) $Adu \cdot \alpha(e_{ij}) = e_{j+1,i+1}$ (respectively $Adu \cdot \alpha(e_{ij}) = e_{i+1,j+1}$) ;

(*b'*) $Adu \cdot \alpha(e_{ij}) = e_{ji}$ if $\alpha \in Ant(M)$)

c) $\|u - \mathbf{I}\|_\psi \leq \varepsilon;$

d) $u\alpha(u^*)\alpha^2(u)...\alpha^{p-1}(u^*) = \mathbf{I}$

(respectively $u\alpha(u)\alpha^2(u)...\alpha^{p-1}(u) = \mathbf{1}$).

Proof. Using the Proposition 2.3.3 and following the proofs of Lemmas 2.1 and 5.7.1 in [G 3] we can construct a system of matrix units $\{(F_{ij}), i, j = 1, ..., p\}$ of M_ω (respectively, in $\Re_\omega$), such that

$$\alpha_\omega(F_{ij}) = F_{j+1,i+1} \quad (\text{respectively, } \alpha_\omega(F_{ij}) = F_{i+1,j+1}).$$

Let $\{(f_{ij}^k), i, j = 1, ..., p\}_{k\in\mathbb{N}}$ be the sequence of the systems of matrix units of M (respectively, of $\Re$), which is represented by $\{(F_{ij}), i, j = 1, ..., p\}$ (see [D 1, p. 31], [S 5, Lemma 2.8]). Since the projection $f_{2,2}^k$ is equivalent to $\alpha(f_{2,2}^k)$ and $\alpha_\omega(F_{1,1}) = F_{2,2}$, by Lemma 1.1.4 of [C 2] there exists a sequence of the partial isometries $(u_k)_{k\in\mathbb{N}}$ in M (respectively, in $\Re$), such that

$$u_k u_k^* = \alpha(f_{1,1}^k),$$

$$u_k^* u_k = f_{2,2}^k, \ (u_k - f_{2,2}^k) \to 0 \ * -\text{strongly when } k \to \omega \tag{2.1}$$

Put

$$v_k = \sum_{j=1}^{p} \alpha(f_{1,j}^k) u_k f_{2,j+1}^k$$

$$(\text{respectively, } \ v_k = \sum_{j=1}^{p} \alpha(f_{j,1}^k) u_k f_{2,j+1}^k), \ k \in \mathbb{N}.$$

It is clear that the element v_k is a unitary in M (respectively, in $\Re$). Moreover, the sequence $(v_k)_{k\in\mathbb{N}}$ is ω-central.

Indeed, according to (2.1) and

$$\sum_{j=1}^{p} \alpha_\omega(F_{1,j} F_{2,2} F_{2,j+1} = \mathbf{1}$$

$$(\text{respectively, } \sum_{j=1}^{p} \alpha_\omega(F_{j,1} F_{2,2} F_{2,j+1} = \mathbf{1})$$

it follows that $(v_k - \mathbf{I}) \to 0$ *-strongly when $k \to \omega$, i.e. the sequence $(v_k)_{k\in\mathrm{N}}$ is ω-central. Since $u_k^*\alpha(f_{1,1}^k)u_k = f_{2,2}^k$, for $1 \le s,t \le p$ we have

$$Adv_k^* \cdot \alpha(f_{s,t}^k) = \sum_{i,j=1}^{p} f_{j+1,2}^k u_k^* \alpha(f_{j,1}^k)\alpha(f_{s,t}^k)\alpha(f_{1,i}^k)u_k f_{2,i+1}^k =$$

$$= f_{t+1,2}^k u_k^* \alpha(f_{1,1}^k) u_k f_{2,s+1}^k = f_{t+1,s+1}^k$$

$$(\text{respectively, } Adv_k^* \cdot \alpha(f_{s,t}^k) = f_{s+1,t+1}^k).$$

Therefore, for fixed k the proof of statements a), b) c) is complete. The proof of the statement b′) follows from Lemma 2.2 of [G 3]. We shall prove the statement d). Let $u_1 = v_k^*$, $e_{ij} = f_{ij}^k$ for fixed k. Consider the element

$$y = u_1\alpha(u_1^*)\alpha^2(u_1)...\alpha^{p-1}(u_1^*)$$

$$(\text{respectively } y = u_1\alpha(u_1)\alpha^2(u_1)...\alpha^{p-1}(u_1)).$$

Let F be a subfactor of M (respectively of $\Re$), generated by the system $\{(e_{ij}),\ i,j = 1,...p\}$. From b) we see that $\beta = Adu_1 \cdot \alpha$ can be restricted on $F' \cap M$ (respectively on $F \cap \Re'$), and we obtain

$$Ady|_F = (Adu_1 \cdot \alpha)^p|_F = id,$$

i.e. $y \in F' \cap M$ (respectively $y \in F \cap \Re'$). Moreover, $\beta(y) = y^*$ and according to c) we have: $\|y - \mathbf{I}\|_\psi \le \varepsilon$. Then applying Proposition 2.3.3 to β and y, we obtain the following

$$y^* = x\beta(x^*)\beta^2(x)...\beta^{p-1}(x^*)$$

$$(\text{respectively } y^* = x\beta(x)\beta^2(x)...\beta^{p-1}(x))$$

for some $x \in F' \cap M$ (respectively $x \in F \cap \Re'$). Put $u = xu_1$. The proof of the statement d) is straightforward. ■

If we apply Lemmas 2.3.5, 2.3.6 and 2.3.9, and the scheme of proof of Proposition 2.3 from [G 3], then we obtain the following Lemma.

Lemma 2.3.10. *There exists a unitary $u \in M$ (respectively $u \in \Re$) and a sequence of system of $p \times p$ matrix units $\{(e_{ij}^\nu), i,j = 1,...,p\}_{\nu\in\mathrm{N}}$ satisfying the following conditions:*

a) $Adu \cdot \alpha(e^{\nu}_{ij}) = e^{\nu}_{j+1,i+1}$ *(respectively* $Adu \cdot \alpha(e^{\nu}_{ij}) = e^{\nu}_{i+1,j+1}$*);*

(a') $Adu \cdot \alpha(e^{\nu}_{ij}) = e^{\nu}_{ji}$ *if* $\alpha \in Ant(M)$);

b) *Let* K_ν *be the subfactor generated by* $\{(e^{\nu}_{ij}),\ i,j = 1,...,p\}_{\nu\in\mathbb{N}}$, *and* K *(respectively* $\mathcal{K}$*) the factor generated by* $K_\nu, \nu \in \mathbb{N}$*; then* K *(respectively* $\mathcal{K}$*) is factor of type* II_1 *and* $M = K \otimes K'_M$ *(respectively* $\mathfrak{R} = \mathcal{K} \otimes \mathcal{K}'_{\mathfrak{R}}$*);*

c)

$$u\alpha(u^*)\alpha^2(u)...\alpha^{p-1}(u^*) = \mathbf{1}$$

$$(\text{respectively } \ u\alpha(u)\alpha^2(u)...\alpha^{p-1}(u) = \mathbf{1}).$$

Recall that a W^*-algebra M is said to be *hyperfinite*, if M is approximately finite dimensional (see Section 1.6). Similarly we shall define the hyperfiniteness of real W^*-algebra: a real W^*-algebra $\mathfrak{R}$ is said to be *hyperfinite*, if $\mathfrak{R}$ is generated by an increasing sequence of finite dimensional real W^*-subalgebras (see [S 5]).

Now we shall consider examples of minimally and purely periodic *-automorphism and *-antiautomorphism of real and complex hyperfinite factors of type II_1.

Example 2.3.11. a). Let M be a hyperfinite factor of type II_1, and let M_n be the algebra $n \times n$ matrices over the field $\mathbb{C}$ with matrix units $\{e_{ij},\ i,j = 1,...,n\}$. It is known that M can be identified with the infinite tensor product of the couples (M_n, tr_n), where tr_n is the canonical trace on M_n (see [C 5]). Put

$$t_n(e_{ij}) = e_{ji}, \quad \alpha_n(e_{ij}) = e_{j+1,i+1}, \quad \theta_n(e_{ij}) = e_{i+1,j+1}, \quad i,j = 1,...,n.$$

Note that if n is even, then the maps t_n and α_n are periodic *-antiautomorphisms of M_n. Put

$$S_n = \otimes_\infty \alpha_n, \quad t = \otimes_\infty t_n, \quad S^1_n = \otimes_\infty \theta_n.$$

It is easy to see that the map S_n (respectively S^1_n) is purely periodic *-antiautomorphism (respectively minimally periodic *-automorphism) of the hyperfinite factor M and the map t is an involutive *-antiautomorphism of M.

b). Let $\Re$ be a real hyperfinite factor of type II_1 and let M_n be the algebra of $n \times n$ matrices over the field $\mathbf{R}$ with matrix units $\{e_{ij},\ i,j = 1,...,n\}$. Then we can identify $\Re$ with the infinite tensor product of the couples (M_n, tr_n), where tr_n is the canonical trace on M_n. Put $S_n^1 = \otimes_\infty \theta_n$, where $\theta_n(e_{ij}) = e_{i+1,j+1}$, $i,j = 1,...,n$. The map S_n^1 is minimally periodic *-automorphism of hyperfinite real factor $\Re$ (for details see [C 5], [R 2]).

Below let M be an injective factor, $\Re$ - an injective real factor.

Lemma 2.3.12. *Let α be a purely periodic *-antiautomorphism of M (respectively minimally periodic *-automorphism of $\Re$) with period p. Then*

*(i) the *-antiautomorphisms α and $S_p \otimes \alpha$ are conjugate (respectively the *-automorphisms α and $S_p^1 \otimes \alpha$ are conjugate);*

*(ii) if α is a *-antiautomorphism of M, then the *-antiautomorphisms α and $t \otimes \alpha$ are conjugate.*

Proof. By Lemma 2.3.10 we have

1) $M = K \otimes K'_M$ $\quad (\Re = \mathcal{K} \otimes \mathcal{K}'_\Re)$;

2) $u\alpha(u^*)\alpha^2(u)...\alpha^{p-1}(u^*) = \mathbf{1}$ (respectively $u\alpha(u)\alpha^2(u)...\alpha^{p-1}(u) = \mathbf{1}$).

3) $Adu \cdot \alpha(e_{ij}^\nu) = e_{j+1,i+1}^\nu$ (respectively $Adu \cdot \alpha(e_{ij}^\nu) = e_{i+1,j+1}^\nu$), for $\nu \in \mathrm{N}$.

It is clear that

$$(Adu \cdot \alpha)|_K = S_p$$

$$(\text{respectively } (Adu \cdot \alpha)|_{\mathcal{K}} = S_p^1).$$

According to $S_p \sim S_p \otimes S_p$ (respectively to $S_p^1 \sim S_p^1 \otimes S_p^1$) we obtain

$$\beta = S_p \otimes \beta|_{K'} \sim S_p \otimes S_p \otimes \beta|_{K'} = S_p \otimes \beta$$

$$\text{(respectively} \quad \beta \sim S_p^1 \otimes \beta)$$

where $\beta = Adu{\cdot}\alpha$. Similarly, by Lemma 2.3.10 (a') we have $(Adu{\cdot}\alpha)|_K = t$). According to $t \sim t \otimes t$ we obtain $\beta \sim t \otimes \beta$. By 2) Proposition 2.3.2 implies that $Adu \cdot \alpha \sim \alpha$. Hence we obtain

$$\alpha \sim \beta \sim S_p \otimes \beta \sim S_p \otimes \alpha$$

$$\text{(respectively} \quad \alpha \sim S_p^1 \otimes \alpha \qquad \text{and} \qquad \alpha \sim t \otimes \alpha). \quad \blacksquare$$

Similarly to Lemmas 2.3.10 and 2.3.12 we can prove the following Lemma.

Lemma 2.3.13. *Let α be a purely periodic *-antiautomorphism of M (respectively a minimally periodic *-automorphism of $\mathfrak{R}$) with period p. Let v be a unitary in M (respectively a unitary in $\mathfrak{R}$) such that*

$$vS_p(v^*)S_p^2(v)...S_p^{p-1}(v^*) = -\mathbf{I}$$

$$\text{(respectively } vS_p^1(v)(S_p^1)^2(v)...(S_p^1)^{p-1}(v) = -\mathbf{I}).$$

Then

$$\alpha \sim Adv \cdot S_p \otimes \alpha \qquad \text{(respectively} \quad \alpha \sim Adv \cdot S_p^1 \otimes \alpha)$$

Proposition 2.3.14. *Let α and β be purely periodic *-antiautomorphisms of M (respectively minimally periodic *-automorphisms of $\mathfrak{R}$) with period p. If α and β are outer conjugate, then they are conjugate.*

Proof. Since α and β are outer conjugate, $\beta = Adu \cdot \alpha$ for some unitary u. It is easy to see that

$$w := u\alpha(u^*)\alpha^2(u)...\alpha^{p-1}(u^*) = \pm\mathbf{I}$$

$$\text{(respectively} \quad w := u\alpha(u)\alpha^2(u)...\alpha^{p-1}(u) = \pm\mathbf{I})$$

If $w = \mathbf{I}$, then by Proposition 2.3.2 α and β are conjugate. Let $w = -\mathbf{I}$. Then by Lemma 2.3.13 we have:

$$Adu \cdot \alpha \sim (Adv \cdot S_p) \otimes (Adu \cdot \alpha) = Ad(v \otimes u) \cdot (S_p \otimes \alpha).$$

Put $w = v \otimes u$, $\theta = S_p \otimes \alpha$. Since $w\theta(w^*)\theta^2(w)...\theta^{p-1}(w^*) = \mathbf{I}$, by Proposition 2.3.2 $Adw \cdot \theta$ and θ are conjugate, i.e.,

$$Ad(v \otimes u) \cdot (S_p \otimes \alpha) \sim (S_p \otimes \alpha).$$

Thus from Lemma 2.3.12 we have $S_p \otimes \alpha \sim \alpha$, therefore, $Adu \cdot \alpha$ and α are conjugate. Similarly we prove the assertion for *-automorphism α of $\mathfrak{R}$. ∎

Lemma 2.3.15. *Let α be a purely periodic *-antiautomorphism of M with period p such that $\alpha^2 \in \overline{Int}(M)$, and let $(\psi_j)_{j\geq 1} \in M_*$ with $0 \leq \psi_j \leq \Phi$, for $\Phi \in M_*$, $j \geq 1$. Then there exists a sequence of couples $(K_\nu, v_\nu)_{\nu\geq 1}$, where K_ν is a subfactor of M, and v_ν is a unitary in M, such that*

a) K_ν commutes with K_j, $j < \nu$;

b) K_ν is generated by partial isometries (P_j^ν), $j = 1, ..., p$ and the unitary u_ν such that

$$u_\nu^p = \mathbf{I}, \quad u_\nu P_j^\nu u_\nu^* = P_{j+1}^\nu;$$

c) $\|[\psi_r, u_\nu]\| \leq \delta_\nu$;

d) $\|[\psi_r, P_j^\nu]\| \leq \delta_\nu$, $r < \nu$, $j = 1, ..., p$;

e) $\|(v_\nu - \mathbf{I})v_{\nu-1}...v_1\|_\Phi^\# \leq \varepsilon$, where $\|x\|_\Phi^\# = \Phi(x^*x + xx^*)^{1/2}$;

f) $\{e_{ij}^r = \tilde{u}_r^{i-j} \cdot P_j^r,\ i,j = 1, ..., p\}$ is a system of matrix units in K_r, such that

$$\alpha_\nu(e_{ij}^r) = e_{j+1,i+1}^r, \quad r \geq \nu,$$

where $\alpha_\nu = Ada_\nu \cdot \alpha$, $a_\nu = v_\nu ... v_1$;

h) $\|\psi_k \alpha_\nu^{-2} - \psi_k Ad(u_\nu^2 ... u_1^2)^{-1}\| \leq \varepsilon_\nu$.

Proof of Lemma is carried out easily, similarly to the proof of Lemma 5.7.9 in [G 3].

Proposition 2.3.16. *Let M be an injective factor, α a purely periodic *-antiautomorphism of M with period p, such that $\alpha^2 \in \overline{Int}(M)$. Then there exists an involutive *-antiautomorphism β of M such that the *-antiautomorphisms α and $S_p \otimes \beta$ are conjugate.*

Proof. From Lemma 2.3.15 we have the following five conditions:

1) K is a type II_1 subfactor of M, generated by K_ν, and $M = K \otimes K'_M$.

2) there exists a unitary a in M which is the limit of the sequence $(v_\nu \dots v_1)_{\nu \in \mathrm{N}}$ in the *-strong topology.

3) For $\alpha_\infty = Ada \cdot \alpha$ we have: $\alpha_\infty(e^\nu_{ij}) = e^\nu_{j+1,i+1}$.

4) $\|[\psi_k, e^\nu_{ij}]\| \leq 2\delta_\nu, \; k < \nu$.

5) $\alpha^2_\infty|_{K'_M} = id.$

Then according to 1) - 5) we obtain

$$Ada \cdot \alpha = \alpha_\infty|_K \otimes \alpha_\infty|_{K'_M} \sim S_p \otimes \beta,$$

for some involutive *-antiautomorphism β of M. Therefore, α_∞ is purely periodic with the period p. Then by Proposition 2.3.14 the *-antiautomorphisms α and $Ada \cdot \alpha$ are conjugate. Hence, $\alpha \sim S_p \otimes \beta$. ■

Theorem 2.3.17. *Let M be an injective factor of type II_1, II_∞ or III_1. Then the conjugacy class of purely periodic *-antiautomorphisms with the same period is unique.*

Proof. Since for injective factors of type II_1, II_∞ and III_1, we have that $Aut(M) = \overline{Int}(M)$, thus $\alpha^2 \in \overline{Int}(M)$ for all $\alpha \in Aut(M)$. By Theorem 1.6.4 any two involutive *-antiautomorphisms of such factors are conjugate. Therefore, by Proposition 2.3.16 all purely periodic *-antiautomorphisms with the same period are conjugate too. ■

Let $\mathfrak{R}$ be a real injective factor of type II_1. For outer conjugacy classes A and B in $Aut(\mathfrak{R})$, denote by $A \times B$ the outer conjugacy class of any *-automorphism $\alpha \otimes \beta \in Aut(\mathfrak{R} \otimes \mathfrak{R})$ with $\alpha \in A$ and $\beta \in B$. For every $p \in \mathrm{N}$, let Br_p be the outer conjugacy class of *-automorphisms of $\mathfrak{R}$ with outer period p.

Proposition 2.3.18. *For each $p \in \mathrm{N}$ the set Br_p of outer conjugacy classes in $Aut(\mathfrak{R})$, with outer period equal to p, endowed with the composition law $(A, B) \to A \times B$ is an abelian group and $\gamma(\alpha)$ is an isomorphism of this group on the group $\{1, -1\}$.*

Proof. We shall show that if $A, B \in Br_p$, then $A \times B \in Br_p$. Indeed, by Corollary 6 from [C 5] we know that the tensor product $\alpha \otimes \beta$ of two *-automorphisms α and β is inner if and only if both are inner. It is easy to see that $p_0(\alpha \otimes \beta)$ equals to the least commom multiple $(p_0(\alpha), p_0(\beta))$, therefore, $A \times B \in Br_p$. Next we show that the class E of S_p^1 is a unit in Br_p. Indeed, let $A \in Br_p$, and α a minimal periodic *-automorphism from A. Then by Lemma 2.3.12 (i) we know that $\alpha \otimes S_p^1$ is conjugate to α and hence $A \times E = A$. Let $A, B \in Br_p$, $\alpha \in A, \beta \in B$, $\alpha^p = Adu$, $\alpha(u) = \gamma(\alpha)u$, $\gamma(\alpha) = \pm 1$ and $\beta^p = Adv$, $\beta(v) = \gamma(\beta)v$, $\gamma(\beta) = \pm 1$. Then

$$(\alpha \otimes \beta)^p = Ad(u \otimes v), \quad (\alpha \otimes \beta)(u \otimes v) = \gamma(\alpha) \cdot \gamma(\beta)\,(u \otimes v),$$

therefore, γ is a homomorphism. Let $A \in Br_p$, $\alpha \in A$, and let α be minimal periodic with $\gamma(\alpha) = 1$. Then α is purely periodic with period p, therefore, α conjugate to S_p^1, i.e. $A = E$. Let $A \in Br_p$. Since $\gamma(A \times A) = 1$, i.e. $A \times A = E$, then $A^{-1} = A$. It is clear that if $A \neq E$, then for every $\alpha \in A$ we have: $\gamma(\alpha) = -1$. Thus, γ is a homomorphism with trivial kernel and γ is surjective, i.e., γ is an isomophism from Br_p onto $\{1; -1\}$. ■

Corollary 2.3.19. *Let α and β be *-automorphisms of $\mathfrak{R}$ such that $p_0(\alpha), p_0(\beta) > 0$. Then the following condition are equivalent:*

*1) *-automorphisms α and β are outer conjugate;*

2) $p_0(\alpha) = p_0(\beta)$, $\gamma(\alpha) = \gamma(\beta)$.

Theorem 2.3.20. *Let $\mathfrak{R}$ be a real injective factor of type II_1, α and β periodic *-automorphisms of $\mathfrak{R}$. Then the following condition are equivalent:*

*a) *-automorphisms α and β are conjugate;*

*b) *-automorphisms α and β are outer conjugate, and the inner *-automorphisms $\alpha^{p_m(\alpha)}$ and $\beta^{p_m(\beta)}$ are conjugate.*

Proof. The assertion a) $\Rightarrow$ b) is trivial. We shall prove the converse assertion. Let *-automorphisms α and β be outer conjugate. Then by Corollary 2.3.19 we have $p_m(\alpha) = p_m(\beta) = n$ for some $n \in \mathbb{N}$. Let u and v be unitaries in $\mathfrak{R}$ such that

$$\alpha^n = Adu, \ \alpha(u) = u, \quad \text{and} \quad \beta^n = Adv, \ \beta(v) = v.$$

Since the *-automorphism α is periodic, u has a finite spectral decomposition:

$$u = \sum_{j=1}^{n} \lambda_i e_i \in \mathcal{U}(\mathfrak{R}).$$

Put $u^{1/n} = \sum \lambda_i^{1/n} e_i$. It is easy to see that $u^{1/n}$ is a unitary, and $u^{1/n} \in \mathfrak{R}^\alpha = \{x \in \mathfrak{R} : \alpha(x) = x\}$. Indeed, let α_1 be an involutive *-antiautomorphism of $\mathcal{U}(\mathfrak{R})$ generating $\mathfrak{R}$. Since $u \in \mathfrak{R}^\alpha$, we have that $\alpha_1(e_i) = e_{-i}$, for all i. Then we check directly that $\alpha_1(u^{1/n}) = (u^{1/n})^*$, i.e. $u^{1/n} \in \mathfrak{R}^\alpha$.

Put

$$\tilde{\alpha} = Adu^{-1/n} \cdot \alpha.$$

Similarly, we shall construct $\tilde{\beta} = Adv^{-1/n} \cdot \beta$. It is clear that the *-automorphisms $\tilde{\alpha}$ and $\tilde{\beta}$ are minimally periodic, and by condition of theorem they are outer conjugate. By Proposition 2.3.14 they are conjugate, i.e., there exists a *-automorphism θ of $\mathfrak{R}$ such that $\tilde{\beta} = \theta \cdot \alpha \cdot \theta^{-1}$. Thus

$$\theta \cdot \alpha \cdot \theta^{-1} = Ad\theta(u^{1/n}) \cdot \tilde{\beta}. \tag{2.2}$$

Now we shall construct a *-automorphism θ_1 of $\mathfrak{R}$, commuting with $\tilde{\beta}$ and such that $\theta_1\theta(u^{1/n}) = v^{1/n}$. It is clear that $v^{1/n} \in \mathfrak{R}^{\tilde{\beta}}$, and (2.2) implies $\theta(u^{1/n}) \in \Re^{\tilde{\beta}}$. Note that $e_j \in \mathcal{U}(\Re^{\alpha})$, $f_j \in \mathcal{U}(\Re^{\beta})$ (see [C 5, 2.2 and 2.4]). Since the *-automorphisms Adu and Adv are conjugate, projections e_j and f_j are equivalent in $\Re$, for each $j = 1, ..., n$.

Therefore, $\tau(e_j) = \tau(f_j)$, where τ is the canonical trace on $\mathcal{U}(\Re)$. Then according to $\tau(\theta(e_j)) = \tau(f_j)$, and since $\theta(e_j), f_j \in \mathcal{U}(\Re^{\tilde{\beta}})$ it's clear that the projection $\theta(e_j)$ is equivalent to f_j in $\Re^{\tilde{\beta}}$. Therefore there exists a partial isometry u_j in $\Re^{\tilde{\beta}}$ such that $u_j\theta(e_j)u_j^* = f_j$, for every $j = 1, ..., n$. Put $w = \sum_{j=1}^{n} u_j$. It is clear that the element w is a unitary in $\Re^{\tilde{\beta}}$. By construction we have:

$$Adw \cdot \theta(u^{1/n}) = v^{1/n}.$$

Checked directly that the *-automorphism $\theta_1 = Adw$ commutes with $\tilde{\beta}$. Then by (2) we obtain: $(Adw \cdot \theta) \cdot \alpha \cdot (Adw \cdot \theta)^{-1} = \beta$, therefore, the *-automorphisms α and β are conjugate. ∎

Let R be a hyperfinite factor of type II_1, $\Re$ a real hyperfinite factor of type II_1. Let $\theta = Adw$ be a periodic inner *-automorphism of R (respectively of $\Re$) with period n. Consider

$$w = \sum_{j=1}^{n} \lambda_k e_k$$

the spectral decomposition of w in R (respectively in $\mathcal{U}(\Re)$). We define now the inner invariant $\varepsilon(\theta)$ to be the probability measure

$$\sum_{j=1}^{n} \tau(e_k)\varepsilon_{\lambda_k},$$

determined up to a rotation on $T = \{z \in \mathbf{C},\ |z| = 1\}$, where ε_{λ_k} is the measure, concentrated on $\lambda_k \in T$, $k = 1, ..., n$.

It is easy to see that two periodic inner *-automorphisms Adu and Adv are conjugate if and only if $\varepsilon(Adu) = \varepsilon(Adv)$.

Theorem 2.3.21. *Let R be a hyperfinite factor of type II_1, α and β periodic *-antiautomorphisms of R. Then the *-antiautomorphisms α and β are conjugate if and only if $p_0(\alpha) = p_0(\beta)$, and $\varepsilon(\alpha) = \varepsilon(\beta)$.*

Proof. The condition is clearly necessary. We shall prove the converse assertion. Let $p_0(\alpha) = p_0(\beta) = m$. Then

$$\alpha^m = Adu, \; u = \sum_{j=1}^{m} \lambda_j e_j, \; \beta^m = Adv, v = \sum_{j=1}^{m} \lambda_j f_j.$$

Put

$$\tilde{\alpha} = Adu^{-1/m} \cdot \alpha, \quad \tilde{\beta} = Adv^{-1/m} \cdot \beta.$$

Checked directly that the *-antiautomorphisms $\tilde{\alpha}$ and $\tilde{\beta}$ are purely periodic with the period m. Then by Theorem 2.3.17 they are conjugate, i.e. there exists a *-automorphism θ of R such that $\theta\tilde{\alpha}\theta^{-1} = \tilde{\beta}$. Since

$$\alpha = \tilde{\alpha} \cdot Adu^{1/m} = Adu^{1/m} \cdot \tilde{\alpha},$$

it follows that

$$\theta \cdot \alpha \cdot \theta^{-1} = \tilde{\beta} \cdot Ad\theta(u^{1/m}) = Ad\theta(u^{1/m}) \cdot \tilde{\beta}.$$

Similarly to the proof of Theorem 2.3.20 it is easy to construct a *-automorphism Adw of R commuting with $\tilde{\beta}$ and such that

$$Adw \cdot \theta(u^{1/m}) = v^{1/m}.$$

Then we obtain

$$Adw \cdot \theta \cdot \alpha \cdot \theta^{-1} \cdot Adw^* = Adw \cdot Ad\theta(u^{1/m}) \cdot \tilde{\beta} \cdot Adw^* =$$

$$Adw \cdot Adw^* \cdot Adv^{1/m} \cdot Adw \cdot \tilde{\beta} \cdot Adw^* = Adv^{1/m} \cdot \tilde{\beta} = \beta,$$

i.e. $(Adw{\cdot}\theta){\cdot}\alpha{\cdot}(Adw{\cdot}\theta)^{-1} = \beta$, and therefore, the *-antiautomorphisms α and β are conjugate. ∎

Theorem 2.3.22. *Let $\mathfrak{R}$ be a real hyperfinite factor of type II_1, and let α and β be periodic *-automorphisms of $\mathfrak{R}$. Then they are conjugate if and only if $p_0(\alpha) = p_0(\beta)$, $\gamma(\alpha) = \gamma(\beta)$, $\varepsilon(\alpha) = \varepsilon(\beta)$.*

Proof. By Corollary 2.3.19 the *-automorphisms α and β are outer conjugate if and only if $p_0(\alpha) = p_0(\beta)$, $\gamma(\alpha) = \gamma(\beta)$. Since the inner *-automorphisms $\alpha^{p_m(\alpha)}$ and $\beta^{p_m(\beta)}$ are conjugate if and only if $\varepsilon(\alpha) = \varepsilon(\beta)$, then by Theorem 2.3.20 the *-automorphisms α and β are conjugate if and only if $p_0(\alpha) = p_0(\beta)$, $\gamma(\alpha) = \gamma(\beta)$, $\varepsilon(\alpha) = \varepsilon(\beta)$. ■

Let $\mathfrak{R}$ be a real W^*-algebra and α a periodic *-antiautomorphism of $\mathfrak{R}$. It is easy to see that the map $\overline{\alpha}(x) = \alpha(x^*)$ is a *-automorphism of $\mathfrak{R}$. It is clear that two *-antiautomorphisms α and β are conjugate if and only if $\overline{\alpha}$ and $\overline{\beta}$ are conjugate. Hence, by Theorem 2.3.22 we obtain the following

Theorem 2.3.23. *Let $\mathfrak{R}$ be a real hyperfinite factor of type II_1, and let α and β be periodic *-antiautomorphisms of $\mathfrak{R}$. Then they are conjugate if and only if $p_0(\overline{\alpha}) = p_0(\overline{\beta})$, $\gamma(\overline{\alpha}) = \gamma(\overline{\beta})$, $\varepsilon(\overline{\alpha}) = \varepsilon(\overline{\beta})$.*

However, we note that *-antiautomorphisms of real factors essentially differ from those of complex factors, as the following example shows.

Example 2.3.24. Let I be a set of indices and let $(u_n)_{n\in I}$ be a set of unitary elements in $\mathfrak{R}$ such that $u_n^2 = \pm\mathbf{1}$, and the projections e_n^i and e_m^j are not equivalent for $m \neq n$, where e_n^i is the spectral projection of the unitary element u_n, $i, j = 1, 2$, $n \in I$. Then it is clear that for any n and m $(n \neq m)$ the *-automorphisms Adu_n and Adu_m are not conjugate. Let $\alpha_n(x) = Adu_n(x^*)$ for each $n \in I$ and for all $x \in \mathfrak{R}$. Then it is easy to see that for any n and $m (m \neq n)$ the involutive *-antiautomorphisms α_n and α_m are also not conjugate. Note that in the complex case all involutive *-antiautomorphisms are conjugate (see Theorem 1.6.5).

In the following Section we use the above results for the description of outer conjugacy classes of *-automorphisms and *-antiautomorphisms of injective (i.e. hyperfinite) factor of type II_∞.

2.4. Outer conjugacy classes of *-antiautomorphisms and *-automorphisms of complex and real factors.

In this Section we shall consider *-automorphisms of real II_∞ factors and *-antiautomorphisms of complex II_∞ factors. We shall give their classification and prove the uniqueness of the class of *-automorphisms (with $mod(\cdot) = \lambda$) of the injective real II_∞ factor and the uniqueness of the class of *-antiautomorphisms (with $mod(\cdot) = \sqrt{\lambda}$) of the injective complex II_∞ factor.

Let M be a W^*-algebra or a real W^*-algebra, $\theta \in Aut(M)$, and let $p(\theta)$ be the largest projection $e \in M$, $\theta(e) = e$ such that the reduced *-automorphism $\theta^e = \theta|_{eMe}$ is inner. If θ is a *-antiautomorphism of the W^*-algebra M, then put $p(\theta) = p(\theta^2)$. We say that a *-automorphism θ (or a *-antiautomorphism θ) is *properly outer* if $p(\theta) = 0$.

Let $\mathfrak{R}$ be a σ-finite real W^*-algebra, $\mathcal{U} = \mathfrak{R} + i\mathfrak{R}$ the enveloping W^*-algebra for $\mathfrak{R}$. Let α be an involutive *-antiautomorphism of $\mathcal{U}$ generating $\mathfrak{R}$. For $\theta \in Aut(\mathfrak{R})$ let $\overline{\theta} \in Aut(\mathcal{U})$ be the *-automorphism, which is the linear extension of θ on $\mathcal{U}$.

Lemma 2.4.1. $p(\theta) = 0$ *if and only if* $p(\overline{\theta}) = 0$.

Proof. Evidently, if $p(\overline{\theta}) = 0$, then $p(\theta) = 0$. Let $p(\overline{\theta}) \neq 0$. Then there exists a projection e from $\mathcal{U}$ such that $\overline{\theta}(e) = e$, $\overline{\theta}^e = Adw$, $w \in e\mathcal{U}e$. Since $\overline{\theta}(\alpha(e)) = \alpha(e)$ and $\alpha(e), e \in \mathcal{U}^{\overline{\theta}}$, as well as $\alpha(e) \sim e$ in $\mathcal{U}$, we have that $\alpha(e)$ and e are equivalent in $\mathcal{U}^{\overline{\theta}}$. Then there exists a unitary u from $\mathcal{U}^{\overline{\theta}}$ such that $ueu^* = \alpha(e)$ and $\alpha(u) = u$. Consider the *-antiautomorphism $\alpha_1 = Adu \cdot \alpha$. Since $\alpha(u) = u$, *-antiautomorphisms α and α_1 are conjugate. It is obvious that $\alpha_1(e) = e$ and $\alpha_1\overline{\theta} = \overline{\theta}\alpha_1$, since $\overline{\theta}(u) = u$. By Lemma 1.1 in [G 2] there is a unitary v from $\mathcal{U}^{\overline{\theta}}$ such that $u = v\alpha(v)$. Put $\sigma = Adv$. It is clear that $\overline{\theta}\sigma = \sigma\overline{\theta}$.

Denote $f = \sigma^{-1}(e)$. It is easy to see that $\alpha(f) = f$ and $\theta(f) = f$. Put $u_1 = \sigma^{-1}(w)$. By a simple calculation we can verify that $\theta^f = Adu_1$, $u_1^* u_1 = u_1 u_1^* = f$. Hence, $p(\theta) \neq 0$, $p(\theta) \geq f$, $f \neq 0$. ∎

Lemma 2.4.2. *If $\theta \notin Int(\Re)$, then for any $\varepsilon > 0$ there exists a projection $e \neq 0$ such that $\|e\theta(e)\| \leq \varepsilon$.*

Proof of Lemma is carried out easily, similarly to the proof of Lemma 1.2.3 in [C 2].

We are going to obtain a similar statement for *-antiautomorphisms.

Let M be a σ-finite W^*-algebra and $\theta \in Ant(M)$.

Lemma 2.4.3. *If $\theta^2 \notin Int(M)$, then for any $\varepsilon > 0$ there exists a projection $f \in M, f \neq 0$, such that $\|f\theta(f)\| \leq \varepsilon$.*

Proof. Let β be an involutive *-antiautomorphism of M, such that $\theta\beta = \beta\theta$ (see [R 2, Lemma 1.4.1]). Put $\gamma = \theta\beta$. Let $M^\beta = \{x \in M : \beta(x) = x^*\}$ be the real W^*-algebra corresponding to β. We apply Lemma 2.4.2 to the *-automorphism γ, and obtain $\|f\gamma(f)\| \leq \varepsilon$, where f is an appropriate projection in M^β. Therefore, $\|f\gamma(f)\| = \|f\theta\beta(f)\| = \|f\theta(f)\| \leq \varepsilon$. ∎

Now we shall prove an analogue of Rokhlin Lemma which differs from the complex case (see [C 2, Theorem 1.2.1]).

Lemma 2.4.4.

(A) Let $\Re$ be a σ-finite real W^*-algebra, $\theta \in Aut(\Re)$.

(i) If θ is properly outer then

$$\left.\begin{array}{ll} \text{for any nonzero projection } e \in \Re \text{ and any} & \varepsilon > 0, \\ \text{there exists a nonzero projection } f \leq e \text{ such that} & \|f\theta(f)\| \leq \varepsilon. \end{array}\right\} \quad (*)$$

(ii) Let θ satisfy the condition $(*)$. If there is no projection $g \in \Re$ such that $\theta^g = Adv$ and $Spv = \{\pm i\}$, then θ is properly outer.

(B) Let M be a σ-finite W^*-algebra, $\theta \in Ant(M)$.

(iii) If θ is properly outer then it satisfies the condition $(*)$.

(iv) Let θ satisfy the condition $(*)$. If there is no projection $g \in M$ such that $(\theta^2)^g = Adv$ and $Spv = \{\pm i\}$, then θ is properly outer.

Proof. Proofs of the statements (i) and (iii) are carried out easily, similarly to the Proof of Theorem 1.2.1 in [C 2], if we apply Lemmas 2.4.2 and 2.4.3.

Proof of statement (ii). Let $\theta \in Aut(\Re)$, and assume first that $p(\theta) \neq 0$. Then we may suppose without loss of generality that $\theta = Adu$, $Spu \neq \{\pm i\}$, where u is a unitary in $\Re$. Let e be a spectral projection of u such that $\|ue - \lambda e\| \leq \varepsilon/2$, where $\lambda \in \mathbf{C}$, $|\lambda| = 1$. In general, we know that $e \in \mathcal{U}$. Suppose that there is no projection e_1 in $\Re$ such that $e_1 \leq e$, and $e_1 \neq 0$. Then the projections e and $\alpha(e)$ are orthogonal, and $e = \frac{1}{2}f + iq$, where $f = \alpha(e) + e$ and $q \in \Re$, $q^* = -q$, $q^*q = -q^2 = \frac{1}{4}f$. We obtain easily that $\|u\alpha(e) - \overline{\lambda}\alpha(e)\| \leq \frac{\varepsilon}{2}$ too. Thus

$$\|Ad(uf) - Ad(\lambda e + \overline{\lambda}\alpha(e))\| \leq \varepsilon,$$

i.e. $\|\theta^f - Adw\| \leq \varepsilon$, where $w = \lambda e + \overline{\lambda}\alpha(e)$ is a unitary from $\Re^f$. Since $\lambda \neq \pm i$, one has that $\lambda = a + ib$, where $a, b \in \mathbf{R}$ and $a \neq 0$. Hence, $w = af - 2bq$.

Let g be a projection from $\Re^f$, $g \neq 0$. Then we obtain easily that

$$\|gAdw(g)\|^2 = \|gAdw(g)g\| \geq \|a^2 g\|.$$

Therefore, $\|gAdw(g)\| \geq |a| > 0$. This is a contradiction with the condition $(*)$.
If there exists a real projection $g \leq e$, $g \neq 0$, then we put $f = g$.

Now we shall give the proof of statement (iv). Let θ be a *-antiautomorphism M, and $p(\theta) \neq 0$. Put $\gamma = \theta\beta$, where β is a *-antiautomorphism, commuting with θ. Since the *-automorphism γ^2 is not properly outer, i.e. $p(\gamma^2) \neq 0$, then $p(\gamma) \neq 0$. Let $M^\beta = \{x \in M : \beta(x) = x^*\}$. By Lemma 2.4.1 the *-automorphism $\theta_1 = \gamma|_{M^\beta}$ is not properly outer. As in the proof of statement (ii), we shall construct a projection $e \in M^\beta$ such that for each projection $f \neq 0$ with $f \leq e$ we have $\|\theta_1(f)f\| \geq \frac{1}{2}$. Then we obtain

$$\|\theta_1(f)f\| = \|\theta(\beta(f))f\| = \|\theta(f)f\| \geq \frac{1}{2}.$$

Hence, we have constructed a projection e such that $\|f\theta(f)\| > \varepsilon$ for each projection $f \neq 0$ with $f \leq e$. We obtain a contradiction. ■

Remark 1. In the discrete case the condition (*) does not imply that θ is properly outer.

Indeed, let $\Re = M_2(\mathbf{R})$ and $\theta = Adu$, where u is the unitary defined as

$$u = \begin{pmatrix} 0 & 1 \\ -1 & 0 \end{pmatrix}$$

Let f be an arbitrary projection from $\Re$. Then f has the following representation:

$$f = \begin{pmatrix} a & b \\ b & 1-a \end{pmatrix} \text{ if } b \neq 0, \text{ and } f = \begin{pmatrix} a & 0 \\ 0 & c \end{pmatrix} \text{ if } b = 0.$$

It is easy to check that if $f \neq \mathbf{1}$ then $f\theta(f) = 0$.

Remark 2. If $\Re$ is a continuous σ-finite real W^*-algebra, then the condition $(*)$ implies that θ is properly outer.

In fact let us have the condition $(*)$ and suppose that $p(\theta) \neq 0$. By Lemma 2.4.1 it follows that $p(\overline{\theta}) \neq 0$ and there is a unitary $u \in \Re$ such that $\theta = Adu$. Then from [C 2] we know that for an appropriate spectral projection $e \in M$ of u we have

$$\|e_1\overline{\theta}(e_1)\| \geq 1/2 \quad \text{for every} \quad e_1 \leq e, \ e_1 \neq 0.$$

Let e be an arbitrary spectral projection of u. Consider the subalgebra $\mathcal{U}^{\overline{\theta}} = \Re^\theta + i\Re^\theta$. It is clear that $u, e \in \mathcal{U}^{\overline{\theta}}$. One can construct easily a projection $f \in \mathfrak{R}^{\theta}$, such as $f \sim e$. Then $e = wfw^*$, where w is a partial isometry from $\mathcal{U}^{\overline{\theta}}$. Let $e_1 = wf_1w^*$, where f_1 is a subprojection of f and f_1 satisfies the condition $(*)$. Then

$$\|e_1\theta(e_1)\| = \|wf_1w^*\theta(w)\theta(f_1)\theta(w^*)\| =$$
$$= \|wf_1\theta(f_1)w^*\| = \|f_1\theta(f_1)\| \leq \varepsilon.$$

This is a contradiction, and therefore $p(\theta) = 0$.

Remark 3. In the case of complex W^*-algebras a $*$-automorphism θ is properly outer iff it satisfies the condition $(*)$ (see [C 2, Theorem 1.2.1]).

By definition, a $*$-automorphism(respectively, a $*$-antiautomorphism) θ of a real W^*-algebra $\mathfrak{R}$ (respectively, a complex W^*-algebra M) is *aperiodic* iff all its degrees θ^n, $n \neq 0$, are properly outer.

If we apply Lemma 2.4.4 and the scheme of proof of Theorem 1.2.5 [C 2], then we can prove a non-commutative analogue of theorem of Rokhlin for $*$-antiautomorphisms of W^*-algebras and $*$-automorphisms of real W^*-algebras.

Theorem 2.4.5.

1) Let $\mathfrak{R}$ be a finite real W^*-algebra, τ a faithful normal trace on $\mathfrak{R}$, $\tau(\mathbf{1}) = 1$, and suppose that θ is an aperiodic $*$-automorphism of $\mathfrak{R}$ which preserves τ. Then for any integer n and any $\varepsilon > 0$, there exists a partition of unity $\{F_i, i = 1, ..., n\}$ in $\mathfrak{R}$, such that

$$\|\theta(F_1) - F_2\|_2 \leq \varepsilon, \ldots, \|\theta(F_k) - F_{k+1}\|_2 \leq \varepsilon, \ldots, \|\theta(F_n) - F_1\|_2 \leq \varepsilon;$$

2) Let M be a finite W^*-algebra, τ a faithful normal trace on M, $\tau(\mathbf{1}) = 1$ and suppose that θ is an aperiodic $*$-antiautomorphism of M which preserves τ. Then for any integer n and any $\varepsilon > 0$, there exists a partition of unity $\{F_i, i = 1, ..., n\}$ in M, such that

$$\|\theta(F_1) - F_2\|_2 \leq \varepsilon, \ldots, \|\theta(F_k) - F_{k+1}\|_2 \leq \varepsilon, \ldots, \|\theta(F_n) - F_1\|_2 \leq \varepsilon.$$

Let $\mathfrak{R}$ be a real W^*-algebra, $\theta \in Aut(\mathfrak{R})$. We shall define the notion of centrally trivial $*$-automorphism for real W^*-algebras.

A *-automorphism θ of $\mathfrak{R}$ is called *centrally trivial*, if for any centralizing sequence $(x_n)_{n\in\mathbf{N}}$ one has:

$$\theta(x_n) - x_n \to 0 \ * -\text{strongly}, \quad \text{when } n \to \infty.$$

The set $Ct(\mathfrak{R})$ of centrally trivial *-automorphisms is a normal subgroup of $Aut(\mathfrak{R})$. It is obvious that if $\theta \in Aut(\mathfrak{R})$, then θ is centrally trivial if and only if $\overline{\theta}$ is centrally trivial in $\mathcal{U}$, where $\overline{\theta}$ is the complexification θ on $\mathcal{U}$.

Now let ω be a free ultrafilter on N, $\mathfrak{R}_\omega$ a real W^*-algebra of central sequences (see Section 2.3), θ_ω a *-automorphism $\mathfrak{R}_\omega$ (if θ is a *-antiautomorphism M, then θ_ω is a *-antiautomorphism M_ω (see [C 2],[C 5])).

Lemma 2.4.6. *Let* $\mathfrak{R}$ *be a real factor with separable predual, and* θ *a *-automorphism of* $\mathfrak{R}$. *Then the following condition are equivalent:*

1) $\theta \notin Ct(\mathfrak{R})$;

2) $\theta_\omega \neq id$;

3) θ_ω *is properly outer.*

The proof immediately follows from Lemma 2.4.6 and Proposition 2.1.2 [C 2].

Let $\mathfrak{R}$ be a real factor, and θ a *-automorphism of $\mathfrak{R}$. *Asymptotic period* of θ is the period of θ modulo $Ct(\mathfrak{R})$; it is denoted by $p_a(\theta)$. In other words $p_a(\theta) \in \mathbf{N}$ and for any $n \in Z$ one has $\theta^n \in Ct(\mathfrak{R})$ iff n is a multiple of $p_a(\theta)$.

For a *-antiautomorphism θ of a W^*-algebra let $p_a(\theta)$ mean the asymptotic period of the *-automorphism θ^2, i.e. $p_a(\theta) = p_a(\theta^2)$.

Lemma 2.4.7. *Let* M *be a real (respectively, complex) factor with separable predual* M_* *and let* θ *be a *-automorphism of* M *(respectively a *-antiautomorphism of* M*) with* $p_a(\theta) = 0$. *Then for any* $n \in \mathbf{N}$ *and any countable subset* $(x^m)_{m\in\mathbf{N}}$ *of* M_ω *there exists a partition of unity* $(f_k)_{k=\overline{1,n}}$ *in* M_ω *such that*

$$(a)\quad \textit{each } f_k \textit{ commutes with all } (x^m)_{m\in\mathrm{N}}\ ;$$

$$(b)\quad \theta_\omega(f_k) = f_{k+1},\quad k=\overline{1,n},\quad f_{n+1}=f_1.$$

Proof. By Lemma 2.4.6 and Theorem 2.4.5 there exists a partition of unity $(\tilde{f}_j)_{k=\overline{1,n}}$ in M_ω, such that

$$\|\theta_\omega(\tilde{f}_j) - \tilde{f}_{j+1}\|_2 < \delta.$$

Let ϕ be a faithful normal state on M, and $(\psi_\nu)_{\nu\in\mathrm{N}}$ a dense sequence in M_*. Then by induction on ν, we use Theorem 2.4.5 and construct a sequence of partitions of unity $(f_k^\nu)_{k=\overline{1,n}}$ in M such that

$$(a)\quad \|\,[\psi_l, f_k^\nu]\,\| \le \tfrac{1}{\nu},\quad l=\overline{1,\nu},\quad k=\overline{1,n};$$

$$(b)\quad \|\,[x_\nu^k, f_j^\nu]\,\| \le \tfrac{1}{\nu},\quad k=\overline{1,\nu},\quad j=\overline{1,n};$$

$$(c)\quad \|\,\theta(f_j^\nu) - f_{j+1}^\nu\,\|_\phi^\sharp \le \tfrac{1}{\nu},\qquad j=\overline{1,n};$$

where $(x_\nu^k)_{\nu\in\mathrm{N}}$ is a representing sequence for (x^k). Let $(f_j)_{j=\overline{1,n}}$ be the collection of partitions of unity in M_ω such that $(f_j^\nu)_{\nu\in\mathrm{N}}$ is a representing sequence for $(f_j)_{j=\overline{1,n}}$. By (b) this collection commutes with all (x^k) and by (c) we have $\theta_\omega(f_j) = f_{j+1}$, $j=\overline{1,n}$. ■

If we use Theorem 2.4.5 and Lemma 2.4.7 and follow the proof of Theorem 2.1.3 in [C 2], we obtain

Theorem 2.4.8.

*1) Let $\Re$ be a real factor with separable predual, θ a *-automorphism of $\Re$ with $p_a(\theta)=0$. Then θ_ω is a stable *-automorphism, i.e. for any unitary u in $\Re_\omega$, there is a unitary $v\in\Re_\omega$ such that $\theta_\omega(v)=uv$.*

*2) Let M be a factor with separable predual, θ a *-antiautomorphism of M with $p_a(\theta)=0$. Then θ_ω is a stable *-antiautomorphism, i.e for any unitary u in M_ω, there is a unitary $v\in M_\omega$ such that $\theta_\omega(v)=uv^*$.*

Let S_p and S_p^1 be the canonical maps constructed in the Example 2.3.11.

Now by helps of Theorem 2.4.8 we shall prove the following Lemma.

Lemma 2.4.9.

1). Let $\Re$ be a real McDuff factor (i.e. $\Re$ is isomorphic to the tensor product $\Re \otimes R$, where R is a real hyperfinite factor of type II_1) with separable predual, and let $\theta \in Aut(\Re)$, $p \in \mathrm{N}$. Then the following conditions are equivalent:

(a) $p_a(\theta) = 0$ (modulo p) ;

(b) $\theta \otimes S_p^1$ *is outer conjugate to* θ;

(c) For any φ *in* $\Re_*^+$, *and any* $\delta > 0$, *there is a unitary* $P \in \Re$ *such that* $\|P - 1\|_\varphi^\# < \delta$ *and* $AdP \cdot \theta$ *is conjugate to* $(AdP \cdot \theta) \otimes S_p^1$

2). Let M be a McDuff factor (i.e. M is isomorphic to the tensor product $M \otimes R$, where R is a hyperfinite factor of type II_1) with separable predual, and let $\theta \in Ant(M)$, $p \in \mathrm{N}$. Then the following conditions are equivalent:

(a) $p_a(\theta) = 0$ (modulo p) ;

(b) $\theta \otimes S_p$ *is outer conjugate to* θ;

(c) For any φ *in* M_*^+, *and any* $\delta > 0$, *there is a unitary* $P \in M$ *such that* $\|P - 1\|_\varphi^\# < \delta$ *and* $AdP \cdot \theta$ *is conjugate to* $(AdP \cdot \theta) \otimes S_p$.

Proof. Since the statements 1) and 2) of this lemma are similar, we shall demonstrate this lemma in parallel way for the both cases.

(a) $\Rightarrow$ (c). Let φ be a faithful normal state on $\Re$ (respectively, on M), and $\{(\psi_j),\ j \in \mathrm{N}\}$ a sequence which is dense in $\Re_*$ (respectively, in M_*). Similar to the proof of Theorem 2.3.1 [C 2], we construct by induction on ν a sequence $\{(P_\nu),\ \nu \in \mathrm{N}\}$ of unitaries in $\Re$ (respectively in M) and a system of matrix units $\{(e_{i,j}^\nu),\ i,j = \overline{1, n_\nu}\}$ in $\Re$ (respectively, in M) which satisfy the following conditions for each $\nu \in \mathrm{N}$:

(1) The real (respectively, complex) factor K_ν, generated by $\{(e^\nu_{i,j}), i,j = \overline{1,n_\nu}\}$, commutes with each $K_1, ..., K_{\nu-1}$;

(2) $\|[e^\nu_{i,j}, \psi_k]\| \leq n_\nu^{-2} 2^{-\nu}$, for $k \leq \nu$ and any $i,j = \overline{1,n_\nu}$;

(3) $P_\nu \in (K_1 \cup ... \cup K_{\nu-1})$;

(4) $\theta_\nu = Ad(P_\nu P_{\nu-1} ... P_1)\,\theta$ satisfies $\theta_\nu(e^k_{i,j}) = e^k_{i+1,j+1}$,

$$\text{(respectively,} \quad \theta_\nu(e^k_{i,j}) = e^k_{j+1,i+1} \; in \; M) \qquad k \leq \nu .$$

(5) $\|(P_\nu P_{\nu-1} ... P_1) - (P_{nu-1} P_{\nu-2} ... P_1)\|^\#_\varphi < \delta 2^{-\nu}$,

where $n_\nu \in N : \; p_a(\theta) = 0 \;\; (mod(n_\nu))$

By (5), $\{(P_{nu} P_{\nu-1} ... P_1), \; \nu \in N\}$ *-strongly converges to a unitary $P \in \Re$ (respectively, $P \in M$), such that $\|P - 1\|^\#_\varphi \leq \delta$. Since $\theta_\nu \longrightarrow \theta_\infty$ when $\nu \longrightarrow \infty$, put $\theta_\infty = AdP \cdot \theta$. By (4), we have:

$$(6) \qquad \theta_\infty(e^k_{i,j}) = e^k_{i+1,j+1} \quad \text{(respectively, } \theta_\infty(e^k_{i,j}) = e^k_{j+1,i+1})$$

for all i,j,k. Combining (2) and Lemma 2.3.6, (respectively, Lemma 2.3.6, [C 2]) we see that K_ν, $\nu \in \mathrm{N}$, generate a type II_1 subfactor $\mathcal{K}$ of $\Re$ (respectively, K of M) which factorizes $\Re$ into $\Re = \mathcal{K} \otimes \mathcal{K}'_\Re$ (respectively, $\mathrm{M} = K \otimes K'$). By (6) the restriction of θ_∞ to $\mathcal{K}$ (respectively, to K) is conjugate to S^1_p (respectively, to S_p). Since S^1_p is conjugate to $S^1_p \otimes S^1_p$ (respectively, S_p is conjugate to $S_p \otimes S_p$) we obtain

$$\theta_\infty = \theta_\infty |_{\mathcal{K}} \otimes \theta_\infty |_{\mathcal{K}'} \sim S^1_p \otimes \theta_\infty |_{\mathcal{K}'} \sim S^1_p \otimes S^1_p \otimes \theta_\infty |_{\mathcal{K}'} \sim S^1_p \otimes \theta_\infty$$

(respectively, $\theta_\infty \sim S_p \otimes \theta_\infty$). This proves (c). The implications (c) $\Rightarrow$ (b) $\Rightarrow$ (a) are obvious. ■

If we use Lemma 2.4.9 and the scheme of proof of Lemma 3.2.7 [C 2], we obtain

Lemma 2.4.10. *Let $\Re$ be a real (respectively,* M *a complex) factor with separable predual, isomorphic to $\Re \otimes R$, where R is a real hyperfinite factor of type II_1 and $\theta \in \overline{Int}(\Re)$ (respectively, isomorphic to $M \otimes R$, where R is a complex hyperfinite factor of type II_1 and $\theta^2 \in \overline{Int}(\mathrm{M})$), with $p_a(\theta) = 0$, and let φ be a faithful normal state on $\Re$ (respectively, on M_*), $\{(\psi_j),\ j \in \mathrm{N}\}$ a sequence of elements of $[0, \varphi]_{\Re_*}$ (respectively, of $[0, \varphi]_{\mathrm{M}_*}$). There exist sequences $\{(K_n),\ n \in \mathrm{N}\}$ of subfactors of $\Re$ (respectively, of M) and $\{(P_n),\ n \in \mathrm{N}\}$ of unitaries in $\Re$ (respectively, in M) such that*

(a) K_n commutes with K_m, $m < n$;

(b) K_n is generated by a partition of unity $\{(F_j^n), j = 1, .., m_\nu\}$

and a unitary U_n, where $U_n^{m_\nu} = \mathbf{I}$,

$$U_n F_j^n U_N^* = F_{j+1}^n, \ j = 1, ..., m_\nu \ ;$$

(c) $\|[\psi_i, U_n]\| \leq \delta_n$ $\|[\psi_i, F_j^n]\| \leq \delta_n$, for all $n \in \mathrm{N}, i < n, j = \overline{1, m_\nu}$;

(d) for any $n \in \mathrm{N}$, P_n commutes with $K_1, ..., K_{n-1}$;

(e) $\|(P_n - 1)P_{n-1}...P_1\|_\varphi^\# \leq 8/m_\nu$;

(f) put $\theta_n = Ad(P_n...P_1) \cdot \theta$, then

$$\theta_n \mid_{K_j} = AdU_j \quad (\text{respectively}, \quad \theta_n^2 \mid_{K_j} = AdU_j), \ j \leq n \ ;$$

(g) for any $j \leq n : \|\psi_j \theta_n^{-1} - \psi_j Ad(U_n U_{n-1}...U_1)^{-1}\| \leq \varepsilon_n$

$$(\text{respectively}, \ \|\psi_j \theta_n^{-2} - \psi_j Ad(U_n U_{n-1}...U_1)^{-1}\| \leq \varepsilon_n).$$

Consider $Out(\Re) = Aut(\Re)/Int(\Re)$. Let ε be the canonical quotient map ε: $Aut(\Re) \longrightarrow Out(\Re)$. Then, with the use of Lemma 2.4.10, similarly to the proof of Theorem 2 [C 2], we easily obtain the following theorem:

Theorem 2.4.11.

1) Let $\Re$ *be a real injective factor with separable predual, and let* θ_1 *and* θ_2 *be *-automorphisms* $\Re$*, such that* $\theta_1, \theta_2 \in \overline{Int}(\Re)$*, with* $p_a(\theta_1) = p_a(\theta_2) = 0$*. Then exists a *-automorphism* $\sigma \in \overline{Int}(\Re)$*, such that* $\varepsilon(\theta_2) = \varepsilon(\sigma\theta_1\sigma^{-1})$*.*

2) Let M *be an injective factor with separable predual, and let* θ_1 *and* θ_2 *be *-antiautomorphisms* M*, such that* $\theta_1^2, \theta_2^2 \in \overline{Int}(M)$*, with* $p_a(\theta_1) = p_a(\theta_2) = 0$*. Then exist a *-automorphism* $\sigma \in \overline{Int}(M)$ *and a unitary* $u \in M$ *such that* $Adu \cdot \theta_2 = \sigma\theta_1\sigma^{-1}$*.*

We shall consider now the set $\mathcal{B}(\Re)$ (respectively, $\mathcal{B}(\mathrm{M})$), of outer conjugacy classes of aperiodic *-automorphisms of $\Re$ (respectively, aperiodic *-antiautomorphisms of M). On $\mathcal{B}(\Re)$ and $\mathcal{B}(M)$ we introduce of the composition $class(\alpha) \cdot class(\beta) = class(\alpha \otimes \beta)$. Let $mod(\cdot)$ be the fundamental homomorphism for $Aut(\Re)$ (respectively, for $Ant(M)$), defined in Section 1.6. Then we use Theorem 2.4.11, and Theorems 2.3.17, 2.3.21 and 2.3.22 on the classification of periodic *-automorphisms and *-antiautomorphisms and prove the following theorem:

Theorem 2.4.12.

1) Let $\Re$ *be a real injective factor of type* II_∞*. Then the map* $mod : \mathcal{B}(\Re) \to [0,\infty)$ *is an isomorphism between the groups* $\mathcal{B}(\Re)$ *and* $[0,\infty)$*.*

2) Let M *be an injective factor of type* II_∞*. Then the map* $mod : \mathcal{B}(M) \to [0,\infty)$ *is an isomorphism between the groups* $\mathcal{B}(M)$ *and* $[0,\infty)$*.*

Here $[0,\infty)$ *is considered as a multiplicative group.*

From the last theorem we obtain the following corollaries.

Corollary 2.4.13. *Let* $\Re$ *be a real injective factor of type* II_∞*. Then for each* $\lambda \in (0,\infty)$*,* $\lambda \neq 1$ *there is a unique (up to outer conjugacy) *-automorphism* θ *of* $\Re$ *such that* $mod(\theta) = \lambda$*.*

Corollary 2.4.14. *Let* M *be an injective factor of type* II_∞*. Then for each* $\lambda \in (0,\infty)$*,* $\lambda \neq 1$*, there is a unique (up to outer conjugacy) *-anti-*

automorphism α *of* M *such that* $mod(\alpha) = \sqrt{\lambda}$.

Theorem 2.1.14 and Corollaries 2.4.13 and 2.4.14 imply

Corollary 2.4.15. *Up to isomorphism there exist exactly two real injective factors of type* III_λ, $\lambda \in (0,1)$.

Note that the existence of exactly two real injective factors of type III_λ, $\lambda \in (0,1)$ was already proved in Theorem 1.7.3 (iii) by an other method.

2.5. Injectivity, amenability, semidiscreteness and hyperfiniteness in real W^*-algebras

In this Section we shall define semidiscreteness and amenability for real W^*-algebra and consider their connection with the injectivity and hyperfiniteness of real W^*-algebras. It is well known that for W^*-algebras the properties of injectivity, amenability, semidiscreteness and hyperfiniteness coincide (see [W], [C 4], [H 2], [EL]). Here we shall consider that for real W^*-algebra these properties are not equivalent in general, namely the hyperfiniteness and the injectivity are not equivalent.

Let $\mathfrak{R}$ and S be W^*-algebras, $\Phi : \mathfrak{R} \to S$ a linear map with $\Phi(\mathbf{1}) = \mathbf{1}$. Denote by $M_n(\mathfrak{R})$ the algebra of $n \times n$ matrix in $\mathfrak{R}$. Suppose that $\Phi_n : M_n(\mathfrak{R}) \to M_n(S)$ is the map defined as

$$\Phi_n\Big([a_{ij}]\Big) = \Big[\Phi(a_{ij})\Big].$$

Similarly to complex W^*-algebras (see [EL]), we consider the following definitions: a maps Φ is called n-**positive**, if the map Φ_n is positive; Φ is **completyly positive**, if it is n-positive for all n; Φ is **morphism**, if it is completely positive and $\Phi(\mathbf{1}) = \mathbf{1}$.

Definition 2.5.1. A real W^*-algebra $\mathfrak{R}$ is called *semidiscrete* if the identity map on $\mathfrak{R}$ can be approximated in the topology of simple-weak* convergence by completely positive normal linear maps of finite rank which preserve the identity.

Proposition 2.5.2. *A real W^*-algebra $\mathfrak{R}$ is semidescrete if and only if the enveloping W^*-algebra $\mathcal{U}(\mathfrak{R})$ is semidiscrete.*

Proof. Let $\mathfrak{R}$ be semidescrete, i.e. there is a sequence of normal morphisms of finite rank $\{\Phi_m\}$, such that $\{\Phi_m\}$ converges to the identity map $I_{\mathfrak{R}}$ *-weakly on $\mathfrak{R}$. Let $I_{\mathcal{U}}$ be the identity map on $\mathcal{U}$. It is clear that $I_{\mathcal{U}} = I_{\mathfrak{R}} + iI_{\mathfrak{R}}$. Then it is obvious that $\{\overline{\Phi}_m = \Phi_m + i\Phi_m\}$ is a sequence of normal morphisms of finite rank and $\overline{\Phi}_m$ *-weakly converges to the identity map $I_{\mathcal{U}}$. Therefore, $\mathcal{U}$ is semidiscrete.

Conversely, let $\mathcal{U}$ be semidiscrete and suppose that $E : \mathcal{U} \to \mathfrak{R}$ is the canonical conditional expectation (see [HS]). It is obvious that (see Lemma 1.3 (iii) from [EL]) the projection E is a normal morphism from $\mathcal{U}$ to $\mathfrak{R}$. Let $\{\overline{\Phi}_m\}$ be the sequence of normal morphisms of finite rank which converges to the identity map $I_{\mathcal{U}}$ *-weakly. Put $\Phi_m = E \circ \overline{\Phi}_m$. It is clear that $\{\Phi_m\}$ is a sequence of normal morphisms of finite rank and $\{\Phi_m\}$ converges to the identity map $I_{\mathfrak{R}}$ *-weakly, i.e. $\mathfrak{R}$ is semidiscrete. ■

Proposition 2.5.3. *If $\{\mathfrak{R}_i\}_{i\in I}$ is a family of real W^*-algebras, then the direct sum $\mathfrak{R} = \oplus_i \mathfrak{R}_i$ is semidiscrete if and only if each $\mathfrak{R}_i$ is semidiscrete.*

Proof. Let ρ be a normal state on $\mathfrak{R}$. Let us prove the necessity. Suppose that $\mathfrak{R}$ is semidiscrete and let $\{\Phi_n\}$ be a sequence of normal morphisms of finite rank which approximates the identity map on $\mathfrak{R}$. For fixed i_0, let Δ_{i_0} be the diagonal morphism from $\mathfrak{R}_{i_0}$ onto $\oplus\mathfrak{R}_{i_0}$ and θ_{i_0} - the projection morphism from $\oplus_i\mathfrak{R}_i$ onto $\mathfrak{R}_{i_0}$. Define the normal morphism $\beta_i : \mathfrak{R}_{i_0} \to \mathfrak{R}_i$ as

$$\beta_i(x) = \begin{cases} x, & \text{if } i = i_0 \\ \rho(x), & \text{if } i \neq i_0 \end{cases}$$

It is easy to see that for any morphism Φ of $\Re$, the map Φ^0 defined as $\Phi^0 = \Delta_{i_0} \cdot (\oplus_i \beta_i) \cdot \Phi \cdot \theta_{i_0} : \Re_{i_0} \to \Re_{i_0}$, is normal morphism of finite rank. Since the sequence $\{\Phi_n\}$ approximates the identity map on $\Re$, it is clear that $\{\Phi_n^0\}$ approximates the identity map on $\Re_{i_0}$.

Conversely, let $\Re_i$ be semidiscrete for any i, and $\{\Phi_n^i\}$ - a sequence of normal morphisms of finite rank which approximates the identity map on $\Re_i$. Let $I_0 \subset I$ and let $\theta : \oplus_{i \in I} \Re_i \to \oplus_{i \in I_0} \Re_i$ be the projection morphism. We denote by h the map defined as $h(x) = x \cdot \mathbf{I}_i$, $h : \Re_i \to \oplus_{i \notin I_0} \Re_i$, where $\mathbf{I}_i$ is the unit of the algebra $\Re_i$. Put

$$\Phi_n = ((\oplus_{i \in I_0} \Phi_n^i) \cdot \theta) \oplus (h \cdot \rho).$$

It is clear that $\{\Phi_n\}$ is a sequence of normal morphisms of finite rank and it is approximates the identity map on $\Re$. ■

Since $\Re$ is isomorphic to $ker(\Phi) \oplus S$ Proposition 2.5.3 implies the following corollary.

Corollary 2.5.4. *If $\Re$ and S are real W^*-algebras, and $\Phi : \Re \to S$ is a normal surjective homomorphism, then $\Re$ is semidiscrete if and only if S and $ker(\Phi)$ are simidiscrete.*

Proposition 2.5.5. *If $\Re$ is a real W^*-algebra, and $\{\pi_i\}_{i \in I}$ is a separating family of normal representations of $\Re$, then $\Re$ is semidiscrete if and only if each of the algebras $\pi_i(\Re)$ is semidiscrete.*

Proof. Let each $\pi_i(\Re)$ be semidiscrete. Suppose that $\{e_k\}_{k \in I}$ is a maximal family of orthogonal central projections such that for each $k \in I$, there is a $i(k) \in I$ for which $\pi_{i(k)}$ is restricted to an isomorphism on $\Re e_k$. Put $S_k = \pi_{i(k)}(\Re e_k)$. By hypothesis the algebra S_k is semidiscrete. Then $\Re$ is isomorphic to $\oplus_{k \in I} S_k$. By Proposition 2.5.3 the algebra $\Re$ is semidiscrete. The converse follows from Corollary 2.5.4. ■

Let $\Re$ and S be real W^*-algebras, and let $\Re \otimes S$ denote their algebraic tensor product, that is each element of $\Re \otimes S$ is an expression of the form $x = \sum_i a_i \otimes b_i$, where $a_i \in \Re$, $b_i \in S$. Let $\mathcal{S}(\Re \otimes S)$ denote the set of all linear functionals f on the vector space $\Re \otimes S$ which satisfy the conditions

$$f(xx^*) \geq 0, \quad f(\mathbf{I}) = \mathbf{I}, \quad \text{for all } x \in \Re \otimes S.$$

For each $f \in \mathcal{S}(\Re \otimes S)$, let $\{\pi_f, \xi_f, H_f\}$ be GNS-representation constructed by f, where ξ_f is a cyclic separating unit vector, and π_f is a homomorphism from $\Re \otimes S$ into the Hilbert space H_f, such that $f(x) =< \pi_f(x)\xi_f, \xi_f >$, $x \in \Re \otimes S$. For each $f \in \mathcal{S}(\Re \otimes S)$, define the seminorm on $\Re \otimes S$ as $\rho_f(x) = \|\pi_f(x)\|$, $x \in \Re \otimes S$. For $\Gamma \subset \mathcal{S}(\Re \otimes S)$ put $\rho_\Gamma = \sup\{\rho_f : f \in \Gamma\}$. If ρ_Γ is a norm, then we call Γ a **separating** subset of $\mathcal{S}(\Re \otimes S)$, and denote by $\mathcal{S}(\Re \otimes_\Gamma S)$ the real W^*-algebra obtained by completing $\Re \otimes S$ with respect to ρ_Γ. Let $\Re^* \otimes S^*$ denote the vector space tensor product of $\Re^*$ and S^*, considered as a space of linear functions on $\Re \otimes S$. We define the following sets in $\mathcal{S}(\Re \otimes S)$:

$$\begin{aligned}
&Min = (\Re^* \otimes S^*) \bigcap \mathcal{S}(\Re \otimes S),\\
&Max = \mathcal{S}(\Re \otimes S),\\
&Bin = \{f \in Max : \text{ the map } (r, s) \to f(r \otimes s)\\
&\text{is separately weak}^* \text{ continuous }\}.
\end{aligned}$$

We denote by ρ_{Min}, ρ_{Max}, and ρ_{Bin} the norms on $\Re \otimes S$ generated by the sets Min, Max, Bin respectively. The completions of $\Re \otimes S$ by those norms we denote respectively by $\Re \otimes_{Min} S$, $\Re \otimes_{Max} S$, and $\Re \otimes_{Bin} S$. It is clear that $Min \subseteq Bin \subseteq Max$, therefore, we have: $\rho_{Min} \leq \rho_{Bin} \leq \rho_{Max}$.

We repeat step by step the scheme of proof of Theorem 4.1 [EL], to obtain the following Theorem:

Theorem 2.5.6. *Let $\Re$ be a real W^*-algebra. Then $\Re$ is semidiscrete if and only if for any real W^*-algebra S,*

$$\Re \otimes_{Min} S \; = \; \Re \otimes_{Bin} S.$$

Let $\Re$ be a real σ-finite W^*-algebra, $\mathcal{U}(\Re)$ its enveloping W^*-algebra, $\alpha_\Re$ - the involutive *-antiautomorphism of $\mathcal{U}(\Re)$ generating $\Re$, let φ be a faithful normal $\alpha_\Re$ - invariant state of $\mathcal{U}(\Re)$. We suppose that $\mathcal{U}(\Re)$ is represented in standart form by φ (see [BR]). Let J_φ be the corresponding modular involution, such that

$$J_\varphi \mathcal{U}(\Re) J_{vp} = \mathcal{U}(\Re)';$$

$$J_\varphi (\mathcal{U}(\Re))' J_{vp} = \mathcal{U}(\Re).$$

We define an involutive *-antiautomorphism $\overline{\alpha}$ on $\mathcal{U}(\Re)'$:

$$\overline{\alpha}(x) = J_\varphi \overline{\alpha}(J_\varphi x J_\varphi) J_\varphi,$$

for every $x \in \mathcal{U}(\Re)'$.

Then the real W^*-algebra

$$\Re' = \{x \in \mathcal{U}(\Re)' : \overline{\alpha}(x) = x^*\}$$

is called the real *commutant* of $\Re$.

It is clear that $\mathcal{U}(\Re') = \mathcal{U}(\Re)'$.

Definition 2.5.7. Let $\Re$ be a real W^*-algebra, and $\eta : \Re \otimes \Re' \to B(H)$ a *-homomorphism defined by:

$$\eta(x \otimes x') = x \cdot x', \quad x \in \Re, x' \in \Re'.$$

$\Re$ is called *amenable*, if η has a bounded extension to $\Re \otimes_{Min} \Re'$.

Theorem 2.5.8. *Let $\Re$ be a real W^*-algebra. Then $\Re$ is amenable if and only if $\Re$ is semidiscrete.*

Proof. Let $\Re$ be semidiscrete. Suppose that $\eta : \Re \otimes \Re' \to B(H)$ is a *-homomorphism defined by $\eta(x \otimes x') = xx'$, $x \in \Re, x' \in \Re'$. Let ω_ξ be the normal state on $B(H)$, corresponding to the unit vector $\xi \in H$. Since $f = \omega_\xi \cdot \eta$ is a linear functional with $f(\mathbf{1}) = 1$, $f(x^* x) \geq 0$ and the map

$(x,x') \to f(x \otimes x')$ is separately weak* continuous (therefore, $f \in Bin$), we have

$$\|\eta(x)\| = \sup_{\|\xi\|=1} [(\omega_\xi \cdot \eta)(x^* x)]^{1/2} \leq \|x\|_{bin}$$

By Theorem 2.5.6 we have $\|\eta(x)\| \leq \|x\|_{min}$, therefore, there exists a *-homomorphism $\overline{\eta}$, extending η to $\mathfrak{R} \otimes \mathfrak{R}'$, i.e. $\mathfrak{R}$ is amenable.

Conversely, let $\mathfrak{R}$ be amenable. Then the *-homomorphism $\eta : \mathfrak{R} \otimes \mathfrak{R}' \to B(H)$ defined as $\eta(x \otimes x') = xx'$, $x \in \mathfrak{R}, x' \in \mathfrak{R}'$, has an extension to $\mathfrak{R} \otimes_{Min} \mathfrak{R}'$, which we denote again by η. Let ξ be the unit vector in H such that the normal state ω_ξ on $B(H)$, corresponding to ξ, has the restriction to $\mathfrak{R}$ and which it is $\alpha_{\mathfrak{R}}$ - invariant (where $\alpha_{\mathfrak{R}}$ is the involutive *-antiautomorphism of $\mathcal{U}(\mathfrak{R})$, generating $\mathfrak{R}$). Then, it is easy to see that the linear functional $f = \omega_\xi \cdot \eta$ is in $Min \cap Bin$. Fix ξ. Denoting by $e' \in \mathfrak{R}'$ the projection with range $\overline{\mathfrak{R}\xi}$, we may spatially identify $\pi_\rho(\mathfrak{R})$ with $\mathfrak{R}e'$. Similarly to the complex case (see Lemmas 4.2 - 4.4 from [EL]) we can construct a completely positive map $T : \mathfrak{R}' \to \mathfrak{R}_*$, corresponding to the state ω_ξ, such that $T(\mathbf{1}) = \rho$ is a state, and also construct the morphism: $\theta_\rho^{-1} \cdot T : \mathfrak{R}' \to e'\mathfrak{R}'e'$ (see [EL, стр 6]). For $x' \in \mathfrak{R}'$ we have:

$$y = \theta_\rho^{-1}(T(x')) \in e'\mathfrak{R}'e'$$

and $T(x')(x) =< xy'\xi, \xi >$, for $x \in \mathfrak{R}$. On the other hand we have:

$$T(x')(x) =< xx'\xi, \xi >=< xe'x'e'\xi, \xi >,$$

and hence $\theta_\rho^{-1} \cdot T(x') = e'r'e'$. Since $Min \cap Bin$ is *-weakly dense in $\mathcal{S}_{Min}(\mathfrak{R} \otimes \mathfrak{R}')$, T is a weakly * limit of the net of weak* continuous completely positive maps $T_\gamma : \mathfrak{R}' \to \mathfrak{R}_*$ with finite rank. From $T(\mathfrak{R}') \subset \mathfrak{R}_*$ it follows that we can assume, that $T_\gamma(\mathbf{1}) = \rho$, therefore, $\{\theta_\rho^{-1} \cdot T_\gamma\}$ is the net of morphisms with finite rank which weak* converges to the morphism $\theta_\rho^{-1} \cdot T$. Now let σ be a normal state on $e'\mathfrak{R}'e'$. We define the net of normal morphisms $\Phi_\gamma : e'\mathfrak{R}'e' \to e'\mathfrak{R}'e'$ as:

$$\Phi_\gamma(x) = \{\theta_\rho^{-1} \cdot T_\gamma(x + \sigma(x)(\mathbf{1} - e'))\}.$$

It is easy to see that the morphism Φ_γ has finite rank and the net $\{\Phi_\gamma\}$ converges to the identity map $I_{e'\mathfrak{R}'e'}$ in the simple weak* topology, i.e. the subalgebra $e'\mathfrak{R}'e'$ is semidiscrete, therefore, $\mathfrak{R}e'$ is semidiscrete. Then by Proposition 2.5.5 $\mathfrak{R}$ is semidiscrete too. ∎

Theorem 2.5.9. *For real W^*-algebras the notions of amenability, semidiscreteness and injectivity are equivalent.*

Proof. Theorem 2.5.8 implies equivalence of amenability and semidiscreteness. Since the notions of amenability, semidiscreteness and injectivity are equivalent in the enveloping W^*-algebra (see [C 4],[H 2]), then by Propositions 2.5.2 and 2.3.8 the proof of Theorem 2.5.9 is complete. ∎

Recall that a real W^*-algebra $\mathfrak{R}$ is said to be *hyperfinite* if there exists an increasing sequence $\{R_n\}$ of finite dimensional real W^*-subalgebras $\{R_n\}$ of $\mathfrak{R}$ with $\mathbf{1} \in R_n$ such that $\bigcup_{n\geq 1} R_n$ is weakly dense in $\mathfrak{R}$ (see Section 2.3).

Proposition 2.5.10. *If a real W^*-algebra $\mathfrak{R}$ is hyperfinite, then the enveloping W^*-algebra $\mathcal{U}(\mathfrak{R})$ is hyperfinite.*

Proof. Let $\mathfrak{R}$ be hyperfinite, i.e. there exists an increasing sequence of finite dimensional real W^*-subalgebras $\{R_n\}$ of $\mathfrak{R}$ with $\mathbf{1} \in R_n$ such that $\bigcup_{n\geq 1} R_n$ is weakly dense in $\mathfrak{R}$. Then it is clear that the sequence $\{\mathcal{U}(R_n)\}$ is increasing and $\bigcup_{n\geq 1} \mathcal{U}(R_n)$ is weakly dense in $\mathcal{U}$, i.e. $\mathcal{U}$ is hyperfinite. ∎

Recall the following two results from [S 5].

Lemma 2.5.11 **[S 5, Lemma 2.2].** *Let $\mathfrak{R}$ be an irreducible real W^*-algebra acting on the n-dimensional Hilbert space $\mathbb{C}^n$, $n < \infty$. Then there exists a conjugate linear isometry J on $\mathbb{C}^n$ such that $JxJ = x$, for all $x \in \mathfrak{R}$, and either $J^2 = \mathbf{1}$, or $J^2 = -\mathbf{1}$. Furthermore if $J^2 = \mathbf{1}$, then $\mathfrak{R}$ is isomorphic to the real $n \times n$ matrices, and if $J^2 = -\mathbf{1}$, then n is* *even and $\mathfrak{R}$ is isomorphic to the $n/2 \times n/2$ matrices over the quaternions.*

Corollary 2.5.12 [S 5, Corollary 2.3]. *Let $\mathfrak{R}_1$ and $\mathfrak{R}_2$ be irreducible real W^*-algebras acting on the finite dimensional Hilbert spaces $\mathbf{C}^{n_1}$ and $\mathbf{C}^{n_2}$ respectively. Let J_k be the conjugate linear isometry on $\mathbf{C}^{n_k}$ associated with $\mathfrak{R}_k$, as in Lemma 2.5.11, $k = 1, 2$. Let $\mathfrak{R}_1 \otimes \mathfrak{R}_2$ be the tensor product of $\mathfrak{R}_1$ and $\mathfrak{R}_2$ acting on $\mathbf{C}^{n_1} \otimes \mathbf{C}^{n_2}$. Then the conjugate linear isometry associated with $\mathfrak{R}_1 \otimes \mathfrak{R}_2$ is $J_1 \otimes J_2$, hence we have:*

(i) if $J_k^2 = \mathbf{1}$ or $J_k^2 = -\mathbf{1}$, for both $k = 1, 2$, then $\mathfrak{R}_1 \otimes \mathfrak{R}_2$ is isomorphic to the real $n_1 n_2 \times n_1 n_2$ matrices;

(ii) if $J_1^2 = \mathbf{1}$ and $J_2^2 = -\mathbf{1}$, or $J_1^2 = -\mathbf{1}$ and $J_2^2 = \mathbf{1}$, then $\mathfrak{R}_1 \otimes \mathfrak{R}_2$ is isomorphic to the $n_1 n_2/2 \times n_1 n_2/2$ matrices over the quaternions.

Proposition 2.5.13. *Let $\mathfrak{R}$ be a properly infinite real factor. Then the following conditions are equivalent:*

(1) $\mathfrak{R}$ is hyperfinite ;

(2) $\mathfrak{R}$ is countably generated, and given elements $x_1, x_2, ..., x_n$ in $\mathfrak{R}$ and any normal state ψ of $\mathfrak{R}$, and $\varepsilon > 0$, there exist a finite dimensional real W^*-subalgebra B of $\mathfrak{R}$ and $y_1, y_2, ..., y_n$ elements in B such that $\|y_k - x_k\|_\psi^\# \leq \varepsilon$, $k = 1, 2, ..., n$;

(3) $\mathfrak{R}$ is the weak closure of the union of an increasing sequence $\{R_n\}$ of real factors with $\mathbf{1} \in R_n$, such that R_n is isomorphic to the real $2^n \times 2^n$ matrices.

Proofs of (1) $\Rightarrow$ (2), and of (3) $\Rightarrow$ (1) are trivial. Proof of (2) $\Rightarrow$ (3) is carried out easily, similarly to the proof of Theorem 3 in [EW]. We shall show that a finite subfactor B, and therefore, subfactors $\{R_n\}$ can be choosed such that they are isomorphic to the real matrices. Indeed, it is clear that a real subfactor B of $\mathfrak{R}$ is isomorphic to an irreducible real factor acting on the 2^m-dimensional complex Hilbert space for some m having the required approximative properties. If B is isomorphic to the real $2^n \times 2^n$ matrices we are through. If not, we can apply Lemma 1.2 in

[S 5] twice to the relative commutant $B' = B' \bigcap \mathfrak{R}$ to find a real subfactor R_4 of B' isomorphic to the real 4×4 matrices, i.e. $R_4 \cong M_4(\mathbf{R})$. By Corollary 2.5.12 we obtain $R_4 = \mathbb{Q} \otimes \mathbb{Q}$, where $\mathbb{Q}$ denotes the quaternions represented as complex 2×2 matrices. If we identify $\mathcal{B} = B \otimes Q \subset \mathfrak{R}$ with a real subfactor of $\mathfrak{R}$, then by Corollary 2.5.12 $\mathcal{B}$ is isomorphic to the real $2^k \times 2^k$ matrices for some k.

Using Proposition 2.5.13, similarly to Corollary 2.10 from [S 5] we can prove the following theorem.

Theorem 2.5.14. *Let M be a properly infinite factor. Then the conjugacy class of involutive *-antiautomorphism generating hyperfinite real factors, is unique.*

Proof. Let α be an involutive *-antiautomorphism of M and let $\mathfrak{R} = \{x \in M \ : \ \alpha(x) = x^*\}$ be the real factor generated by α which is hyperfinite by condition of Theorem. By Proposition 2.5.13 there exists an incresing sequence $\{R_n\}$ of subalgebras of $\mathfrak{R}$, which are isomorphic to the algebra of real $2^n \times 2^n$ matrices, and such that $\mathbf{1} \in R_n$ and $\bigcup_{n \geq 1} R_n$ is weakly dense in $\mathfrak{R}$. Let $M_n = R_n + iR_n$ be the factor of type I_{2^n}. Denote by t_n the transpose map on M_n such that $R_n = \{x \in M_n : t_n(x) = x^*\}$. Since R_n is the algebra of real matrices and the sequence $\{R_n\}$ is weakly approximating $\mathfrak{R}$, then $\alpha(x) = t_n(x)$, for all $x \in M_n$. Thus, any *-antiautomorphism α is the inductive limit of the transpose maps t_n. Since in each matrix algebra all the transpose maps are conjugate by a *-automorphism of $\bigcup_{n \geq 1} M_n$ and since the conjugating *-automorphisms weakly converge to unique *-automorphism of M, then every two involutive *-antiautomorphisms of M generating hyperfinite real factors will be conjugate by a *-automorphism of M. ■

By Corollary 2.4.15 (see also Theorem 1.7.3 (iii)) there exists two classes of non-isomorphic injective real factors of type III_λ, $0 < \lambda < 1$ generating (as the complex enveloping W^*-algebras) the same injective factor of type III_λ. The first of these factors is generated by an involutive *-antiautomorphism with $mod(\alpha)=1$, and the second factor is generated by an involutive *-antiautomorphism with $mod(\alpha)=\sqrt{\lambda}$. We shall construct a hyperfinite real factor $\mathfrak{R}$ of type III_λ, $0<\lambda<1$, with $mod(\alpha_{\mathfrak{R}})=1$ (which we are called *the real Powers factor*).

Let $M_2(\mathbf{R})$ be the algebra of real 2×2 matrices and let φ_λ be the state on $M_2(\mathbf{R})$ defined as

$$\varphi_\lambda\begin{pmatrix} x_{11} & x_{12} \\ x_{21} & x_{22} \end{pmatrix} = \frac{1}{1+\lambda}(\lambda x_{11} + x_{22}).$$

By analogy with the construction of complex Powers factor it is easy to see that if $\lambda=1$, the algebra $\mathfrak{R}_\lambda=\otimes_1^\infty(M_2(\mathbf{R}),\varphi_\lambda)$ is a real factor of type II_1, and if $\lambda=0$ then the algebra $\mathfrak{R}_\lambda$ is a real factor of type I_∞, and when $\lambda\in(0,1)$, the algebra $\mathfrak{R}_\lambda$ is a real factor of type III_λ. Since the state $\omega=\otimes_1^\infty\varphi_\lambda$ on $\mathfrak{R}_\lambda$ satisfies $\sigma_T^{\overline{\omega}}=id$, so by Proposition 2.1.4 (a) and by Remark 2.1.11 we have: $mod(\alpha_{\mathfrak{R}_\lambda})=1$.
Now Theorem 2.5.14 implies the following

Theorem 2.5.15. *A real injective W^*-factor R of type III_λ, $0<\lambda<1$, is hyperfinite if and only if $mod(\alpha_{\mathfrak{R}})=1$.*

Note that a real injective factor S of type III_λ, $0<\lambda<1$ with $mod(\alpha_S)=\sqrt{\lambda}$ can not be approximated by increasing sequence of real or quaternionian matrix subalgebras of S, although its enveloping (complex) factor is approximated by an increasing sequence of complex matrix subalgebras.

We summarize the above results in the following

Theorem 2.5.16. *Let $\mathfrak{R}$ be a real factor and let α be the involutive *-antiautomorphism of $\mathcal{U}(\mathfrak{R})$ generating $\mathfrak{R}$. Then for $\mathfrak{R}$ the following implications are valid:*

Injectivity $\Leftrightarrow$ *Amenability* $\Leftrightarrow$ *Semidiscreteness*

$\Uparrow$ $\Downarrow$ *for* $mod(\alpha_{\mathfrak{R}}) = 1$

H y p e r f i n i t e n e s s

Proof. Theorem 2.5.9 implies the equivalence of properties of injectivity, amenability and semidiscreteness for real factors. By Proposition 2.5.10 the hyperfiniteness implies the injectivity. Theorem 2.5.15 implies that if a real factor $\mathfrak{R}$ is injective and satisfies the condition $mod(\alpha_{\mathfrak{R}}) = 1$, where $\alpha_{\mathfrak{R}}$ is the involutive *-antiautomorphism of $\mathcal{U}(\mathfrak{R})$ generating $\mathfrak{R}$, then $\mathfrak{R}$ is hyperfinite. ∎

2.6. Diameters of state spaces of JW-algebras

In this Section we define the notion of diameter of state spaces for JW-algebras, and calculate it for the case of JW-factors of types I_n $(n < \infty)$, I_∞, II_1, II_∞ and for JW-factors of type III_λ, $0 < \lambda < 1$. The case of JW-factors of type III_0 and III_1 remains open.

First we remind the definition of diameter of the state spaces for W^*-algebras.

Let M be a W^*-algebra, $S_0(M)$ - the set of all normal states on M. For $\omega \in S_0(M)$, we denote by $[\omega]$ the norm closure of the orbit of the state ω under the action of the group $Int(M)$ of all inner *-automorphisms of M given as

$$\omega \rightarrow u\omega u^* = \omega \cdot Adu,$$

where $\omega \cdot Adu = \omega(uxu^*)$, $x \in M$, u being a unitary operator on M. We equip the space $S_0(M)/Int(M)$ of all orbits with the metric

$$d([\omega],[\psi]) = \inf\{\|\omega' - \psi'\| : \omega' \in [\omega]. \ \psi' \in [\psi]\}.$$

The quantity

$$D(M) = \sup\{d([\omega],[\psi]) : [\omega],[\psi] \in S_0(M)/Int(M)\}$$

is called the *diameter of the state space* of the W^*-algebra M.

It is easily seen that if M is not a W^*-factor, then $D(M) = 2$. In [CHS] it is proved that if M is a factor of type I_n $(n < +\infty)$, then $D(M) = 2(1 - 1/n)$ (compare also [Pow]); if M is a W^*-factor of type I_∞, II_1, or II_∞, then $D(M) = 2$; if M is a σ-finite W^*-factor of type III_λ $(0 \leq \lambda \leq 1)$, then $D(M) = 2(1 - \lambda^{1/2})/(1 + \lambda^{1/2})$ (see [CS]).

We define the notion of diameter of state spaces for JW-algebras. Let A be a JW-algebra, $S_0(A)$ - the set of all normal states on A. For $\varphi \in S_0(A)$ we denote by $[\varphi]$ the norm closure of the set of states $\varphi \circ g$, $g \in Int(A)$; that is, the closure of the orbit of φ under the action of the group $Int(A)$. We denote the set of all such classes by $S_0(A)/Int(A)$, and introduce a metric on it, putting

$$d([\omega],[\psi]) = \inf\{\|\omega' - \psi'\| : \omega' \in [\omega],\ \psi' \in [\psi]\}.$$

Definition. We denote by $D(A)$ the diameter of the metric space $S_0(A)$ / $Int(A)$, and call it the *diameter* of the state space of the JW-algebra A.

Proposition 2.6.1. *Let a JW-algebra A be isomorphic to the self-adjoint part of a W^*-algebra M. Then $D(A) = D(M)$.*

Proof. Since $A = M_s = \{x \in M : x = x^*\}$, then, clearly, $S_0(A) = S_0(M)$. From [Up, Theorem 1.4] it follows that every inner *-automorphism of M is the product of a finite number of reflections of the form U_s, where s is a symmetry from $M_s = A$. Therefore $Int(A) = Int(M)$ and hence $D(A) = D(M)$. ■

Proposition 2.6.2. *If a JW-algebra A is not a factor, then $D(A) = 2$.*

Proof. Let e be a central idempotent, $e \neq 0, \mathbf{1}$. For normal states ρ with $\rho(e) > 0$ put $\omega(x) = \rho(e)^{-1}\rho(ex)$, and $\psi(x) = \rho(e^{\perp})^{-1}\rho(e^{\perp}x)$. Then for any $\omega' = \omega \circ g_1$, $\psi' = \psi \circ g_2$, $g_1, g_2 \in Int(A)$, we have

$$\|\omega' - \psi'\| = \sup_{\|x\|\leq 1} \left| \frac{\rho(eg_1(x))}{\rho(e)} - \frac{\rho(e^{\perp}g_1(x))}{\rho(e^{\perp})} \right| \geq$$

$$\geq \left| \frac{\rho(eg_1(e - e^{\perp}))}{\rho(e)} - \frac{\rho(e^{\perp}g_1(e - e^{\perp}))}{\rho(e^{\perp})} \right| =$$

$$= \left| \frac{\rho(e(e - e^{\perp}))}{\rho(e)} - \frac{\rho(e^{\perp}(e - e^{\perp}))}{\rho(e^{\perp})} \right| = \frac{\rho(e)}{\rho(e)} + \frac{\rho(e^{\perp})}{\rho(e^{\perp})} = 2.$$

Here we have used the facts that $\|e - e^{\perp}\| = 1$ and $g_1(e - e^{\perp}) = e - e^{\perp}$, $i = 1, 2$, since e is a central idempotent. Therefore, $d([\omega], [\psi]) = 2$, i.e. $D(A) = 2$. ∎

Thus we can from now on restrict our attention on JW-factors which are not isomorphic to the self-adjoint part of a W^*-algebra.

Theorem 2.6.3. *Let A be a JW-factor of type I_∞ or II. Then*

$$D(A) = 2$$

Proof. Since every JW-factor of type I_∞ or II possesses a faithful normal semifinite trace τ, let p and q be two nonzero idempotents in A, $p \leq q$, with $\tau(p) \leq \tau(q) < +\infty$. Consider two states

$$\omega(x) = \tau(px)/\tau(p), \quad \psi(x) = \tau(qx)/\tau(q)$$

(compare [CHS, p. 92]). Every linear functional φ on A can be extended to a linear functional $\overline{\varphi}$ on $\mathcal{U}(A)$ with the same norm, and, if φ is a trace then so is $\overline{\varphi}$ (see Section 1.2). Since $Int(A) \subseteq Int(\mathcal{U}(A))$, it is easy to see that

$$d([\omega], [\psi]) \geq d([\overline{\omega}], [\overline{\psi}]).$$

Since
$$d([\overline{\omega}],[\overline{\psi}]) = 2(1-\overline{\tau}(p)/\overline{\tau}(q))$$
(see [CHS, p. 92]) then
$$d([\omega],[\psi]) \geq 2(1-\overline{\tau}(p)/\overline{\tau}(q)) = 2(1-\tau(p)/\tau(q)).$$

Now A has type I_∞ or II, so the idempotents p and q may be chosen so that $\tau(p)/\tau(q)$ is arbitrary small. From this it follows that the states ω and ψ may be chosen such that $d([\omega],[\psi])$ is arbitrary close to 2, i.e. $D(A) = 2$. ∎

Thus it remains to examine the cases of JW-factors of types I_n $(n < \infty)$ and III.

Theorem 2.6.4. *For any JW-factor of type I_n $(n < \infty)$,*
$$D(A) = 2(1 - \frac{1}{n}).$$

Proof. By Corollary 1.4.3, every JW-factor of type I_n is isomorphic to one of the following four JW-factors:

1) the algebra $M_n(\mathbf{C})_s$ of self-adjoint $n \times n$ matrices over the field $\mathbf{C}$ of complex number, $n \geq 3$;

2) the algebra $M_n(\mathbf{R})_s$ of self-adjoint $n \times n$ matrices over the field $\mathbf{R}$ of real numbers, $n \geq 3$;

3) the algebra $M_n(\mathbf{Q})_s$ of self-adjoint $n \times n$ matrices over the quaternions, $n \geq 3$;

4) a spin factor (JW-factor of type I_2).

We shall exammine each case.

1) Let $A = M_n(\mathbf{C})_s$. Then from Proposition 2.6.1 it follows at once that
$$D(A) = D(M_n(\mathbf{C})) = 2(1 - 1/n).$$

2) Let $A = M_n(\mathbf{R})_s$. Then A is the self-adjoint part of the real W^*-factor $\Re(A) = M_n(\mathbf{R})$; thus $\mathcal{U}(A) = M_n(\mathbf{C})$.

We shall follow the plan of proof for the case of a W^*-factor of type I_n, $n < \infty$ (see [CHS],[Pow]). Let φ and ψ be normal states of A, and τ be an faithful normal trace on A. By the Radon-Nikodym theorem for JBW-algebras ([A 9], Theorem 8.3 Chap.4) $\varphi(\cdot) = \tau(h\cdot)$, and $\psi(\cdot) = \tau(k\cdot)$, where h and k are elements from A_+. We examine the spectral decomposition of h and k:

$$h = \sum_{i=1}^{n} \lambda_i e_i, \quad k = \sum_{i=1}^{n} \mu_i f_i,$$

where $\{e_i\}$ and $\{f_i\}$ are projections from $\Re(A)$, and $e_i e_j = 0$ and $f_i f_j = 0$ for $i \neq j$; the $\{\lambda_i\}$ and $\{\mu_i\}$ are scalars from $\mathbf{R}_+$.

Let us show that

$$d([\varphi], [\psi]) = \sum_{i=1}^{n} |\lambda_i - \mu_i|.$$

Lemma 2.6.5.

$$\|\varphi - \psi\| \geq \sum_{i=1}^{n} |\lambda_i - \mu_i|.$$

Proof. As we know, φ and ψ always have α_A- invariant extensions $\overline{\varphi}$ and $\overline{\psi}$ on $\mathcal{U}$, preserving the norm: $\|\varphi\| = \|\overline{\varphi}\|$ and $\|\psi\| = \|\overline{\psi}\|$, and such that

$$\overline{\varphi} = \frac{\varphi + \varphi \cdot \alpha_A}{2}, \qquad \overline{\psi} = \frac{\psi + \psi \cdot \alpha_A}{2}.$$

It is easy to see that $\|\varphi - \psi\| = \|\overline{\varphi} - \overline{\psi}\|$.

Let τ be a faithful normal trace on A, and $\overline{\tau} = \tau + \tau \circ \alpha_A$ its α_A-invariant extension on $\mathcal{U} = \mathcal{U}(A)$. Then $\overline{\tau}$ is also a faithful normal finite trace on $\mathcal{U}$ and $\overline{\varphi}(\cdot) = \overline{\tau}(h\cdot)$ and $\overline{\psi}(\cdot) = \overline{\tau}(k\cdot)$. By Theorem 5.6 from [Pow] $\|\varphi - \psi\| = \|\overline{\varphi} - \overline{\psi}\| \geq \sum_{i=1}^{n} |\lambda_i - \mu_i|$. ■

Now we'll show that

$$\|\varphi' - \psi\| \leq \sum_{i=1}^{n} |\lambda_i - \mu_i|$$

for any $\varphi' \in [\varphi]$.
For this we need the following lemma.

Lemma 2.6.6. *Let* $\{e_i\}_{i=1}^n$ *and* $\{f_i\}_{i=1}^n$ *be two families of pairwise orthogonal non zero minimal projections (i.e. atoms) in a JW-factor A of type* I_n, $3 \leq n < \infty$. *Then there exist* $s_1, s_2, ..., s_m$ *symmetries from* A, $m < \infty$ *such that* $U_{s_1} U_{s_2} ... U_{s_m} e_i = f_i$ *for all* $i \in \overline{1,n}$.

Proof. By Theorem 10 and Corollary 17 from [To 1], there exist symmetries $t_1, ..., t_n$ in A such that $U_{t_i} e_i = f_i$. Consider the partial isometries $v_i = f_i t_i e_i \in \Re(A)$, $i \in \overline{1,n}$. Then $v_i e_i v_i^* = f_i$ and $v_i v_j^* = v_i^* v_j = 0$, $i \neq j$. Consider the element $v = \sum_{i=1}^n v_i$. It is a unitary, and $v e_i v^* = f_i$, for all $i \in \overline{1,n}$. Since $v \in \Re(A)$, there exist symmetries $s_1, ..., s_m$ from A, $m < \infty$, such that $Adv = U_{s_1} \cdot ... \cdot U_{s_m}$ [Up]. ∎

We continue the proof for case 2). Let $\{e_i\}_{i=1}^n$ and $\{f_i\}_{i=1}^n$ be spectral projections of the elements h and k respectively. By Lemma 2.6.6 there exist symmetries $s_1, ..., s_m$ from A such that $U_{s_1} \cdot ... \cdot U_{s_m} e_i = f_i$, $i \in \overline{1,n}$. Denote by v the unitary element $v = s_1 \cdot ... \cdot s_m$ in $\Re(A)$. Consider the state

$$\varphi' = \varphi \circ Adv^* = \varphi \circ U_{s_1} \circ ... \circ U_{s_m}.$$

Evidently it is normal, and $\varphi' \in [\varphi]$. Thus

$$\varphi'(x) \;=\; \varphi(v^* x v) \;=\; \tau(h v^* x v) \;=\; \tau(v h v^* x)$$

for all $x \in \Re(A)$, where

$$v h v^* = v \left(\sum_{i=1}^n \lambda_i e_i \right) v^* \;=\; \sum_{i=1}^n \lambda_i v e_i v_i^* \;=\; \sum_{i=1}^n \lambda_i f_i.$$

We calculate the norm of the functional $(\varphi' - \psi)$:

$$||\varphi' - \psi|| \;=\; \sup_{||x|| \leq 1} |\varphi'(x) - \psi(x)| \;=\; \sup_{||x|| \leq 1} |\tau(v h v^* x) -$$

$$-\tau(kx)| \;=\; \sup_{||x|| \leq 1} |\tau((v h v^* - k)x)| \;\leq\; |\tau(v h v^* - k)| \;=$$

$$= \left| \tau \left(\sum_{i=1}^n \lambda_i f_i \;-\; \sum_{i=1}^n \mu_i f_i \right) \right| \;=\; \sum_{i=1}^n |\lambda_i - \mu_i|.$$

Now by Lemma 2.6.5 $\|\varphi'-\psi\| \geq \sum_{i=1}^{n}|\lambda_i-\mu_i|$, so $\|\varphi'-\psi\| = \sum_{i=1}^{n}|\lambda_i-\mu_i|$.

Thus

$$d([\varphi],[\psi]) = \sum_{i=1}^{n}|\lambda_i-\mu_i|.$$

Using this expression for $d([\varphi],[\psi])$, we calculate its greatest value, which is the diameter $D(A)$.

Lemma 2.6.7. *Let* $\lambda_1 \geq \lambda_2 \geq \ldots \geq \lambda_n \geq 0$, $\sum_{i=1}^{n}\lambda_i = 1$; $\mu_1 \geq \mu_2 \geq \ldots \geq \mu_n \geq 0$, $\sum_{i=1}^{n}\mu_i = 1$. *Then*

$$\sup \sum_{i=1}^{n}|\lambda_i-\mu_i| = 2(1 - \frac{1}{n}).$$

Proof. Let us show that $\sum_{i=1}^{n}|\lambda_i-\mu_i| \leq 2(1 - \frac{1}{n})$. Since, for each i,

$$|\lambda_i-\mu_i| = \max(\lambda_i,\mu_i) - \min(\lambda_i,\mu_i)$$

and

$$\max(\lambda_i,\mu_i) = \lambda_i + \mu_i - \min(\lambda_i,\mu_i),$$

we have

$$\sum_{i=1}^{n}|\lambda_i-\mu_i| = \sum_{i=1}^{n}\lambda_i + \sum_{i=1}^{n}\lambda_i - 2\sum_{i=1}^{n}\min(\lambda_i,\mu_i).$$

Now since $\sum_{i=1}^{n}\lambda_i = 1$ and $\sum_{i=1}^{n}\mu_i = 1$, we have $\lambda_1 \geq \frac{1}{n}$, $\mu_1 \geq \frac{1}{n}$, and

$$\sum_{i=1}^{n}\lambda_i + \sum_{i=1}^{n}\mu_i - 2\sum_{i=1}^{n}\min(\lambda_i,\mu_i) \leq$$

$$\leq 2 - 2\min(\lambda_i,\mu_i) \leq 2 - \frac{2}{n} = 2(1-\frac{1}{n}).$$

On the other hand $\sum_{i=1}^{n}|\lambda_i - \mu_i| = 2(1 - \frac{1}{n})$, if $\lambda_1 = 1$ and $\lambda_i = 0, i \in \overline{2,n}$ and $\mu_1 = \mu_2 = ... = \frac{1}{n}$. Therefore

$$\sup \sum_{i=1}^{n}|\lambda_i - \mu_i| = 2(1 - \frac{1}{n}). \quad \blacksquare$$

Thus, for normal states φ_1 and ψ_1 of the form

$$\varphi_1(\cdot) = \tau(e_1 \cdot), \quad \psi_1(\cdot) = \tau\left(\left(\sum_{i=1}^{n}\frac{1}{n}f_i\right)\cdot\right)$$

the distance between the classes $[\varphi_1]$ and $[\psi_1]$ attains greatest value, being equal to the diameter of $S_0(A)/Int(A)$:

$$d([\varphi_1],[\psi_1]) = 2(1 - \frac{1}{n}).$$

The proof of the theorem for case 2) is complete.

3) Let A be isomorphic to the algebra $M_n(\mathbb{Q})_s$ of self-adjoint quaternionian $n \times n$ matrices. The proof for this case proceeds similarly to the case of complex matrices. As we know, every algebra of $n \times n$ matrices over the quaternions can be embedded into the algebra of $2n \times 2n$ matrices over $\mathbb{C}$. Therefore Lemmas 2.6.5 and 2.6.6 are valid also for $M_n(\mathbb{Q})_s$, whence by the scheme of proof for case 2) it is easy to obtain that $D(A) = 2(1 - \frac{1}{n})$.

4) Let A be a spin factor. To each normal state φ on A corresponds its Radon-Nikodym derivative $a \in A_+$ such that $\varphi(\cdot) = \tau(a\cdot)$, τ being a faithful normal trace on A [A 6]. Considering the spin factor in as the direct sum $\mathbf{R} \oplus N$, where N is a real Hilbert space, we obtain $a = (1, x)$ (since φ is a state), and $\|x\| \leq 1$ (since $a \geq 0$). Thus we have one-to-one correspondence between the set $S_0(A)$ of normal states of A and the unit ball U_1 of the space N. It is easy to see that every symmetry s in A has the form $s = (0, y)$, where $\|y\| = 1$.

Let $\varphi \in S_0(A)$ and $\varphi_1 = \varphi \circ U_s$. Then $\varphi_1(\cdot) = \tau(a_1\cdot)$, where $a_1 = U_s = (1, 2(x,y)y - x)$, i.e. $a_1 = (1, x^s)$, where x^s is the vector

$x \in N$ turned through the angle 2α; where α is the angle between x and y.

Thus, applying to φ operators of the form U_s, $s \in A$, $s^2 = \mathbf{I}$, we see that the orbit of $[\varphi]$ consists of all states φ' of the form $\varphi'(\cdot) = \tau(a' \cdot)$, where $a' = (1, z) \in A_+$, $\|z\| = \|x\|$, that is, $[\varphi]$ corresponds to the sphere $S_{\|x\|}$ of the radius $\|x\|$. Therefore $S_0(A)/Int(A) = [0,1]$. Therefore

$$\|\varphi_1 - \varphi_2\| = \|(1, x_1) - (1, x_2)\|_1 = \|(0, x_1 - x_2)\|_1 =$$

$$= \|x_1 - x_2\| \tau(|(0, (x_1 - x_2)/\|x_1 - x_2\|)|) = \|x_1 - x_2\|,$$

since $(0, (x_1 - x_2)/\|x_1 - x_2\|)$ is a symmetry and

$$\tau(|(0, (x_1 - x_2)/\|x_1 - x_2\|)|) = 1.$$

Thus $D(A) = diam[0,1] = 1 = 2(1 - \frac{1}{2})$, if A is a spin factor. Since spin factors are precisely the JW-factors of type I_2, the proof of case 4) is finished. ■

Proposition 2.6.8. *Let A be a reversible JW-algebra. Then the following condition are equivalent:*

(a) A is a σ-finite JW-algebra;

(b) there exist faithful normal state on A;

(c) $\mathcal{U}(A)$ is a σ-finite W^*-algebra.

Proof. a) $\Rightarrow$ b). If ρ is arbitrary normal state on A, then it is clear that there exist the greatest projection $e \in A$ such that $\rho(e) = 0$. If A is σ-finite JW-algebra, then any family of normal states with pairwise orthogonal supports is countable. Let $\{\rho_n\}$ be a maximal family of normal states with supports $\{e_n\}$. Then it is clear that $\sum e_n = \mathbf{I}$. Put

$$\rho(x) = \sum_{n=1}^{\infty} 2^{-n} \rho_n(x), \quad x \in A.$$

Then $\rho(x)$ is a faithful normal state with support $\mathbf{I}$, i.e. ρ is a faithful.

b) $\Rightarrow$ c). Let ρ be a faithful normal state on A. By Corollary 1.2.10 the state ρ can we extended to a faithful normal state on $\mathcal{U} = \mathcal{U}(A)$. By [T 1, Proposition 3.19, Chap. II] $\mathcal{U}(A)$ is a σ-finite W^*-algebra. The statement c) $\Rightarrow$ a) is trivial. ■

Now let A be a σ-finite JW-factor of type III_λ, $(0 < \lambda < 1)$. Then A is reversible, and therefore it is sufficient to consider only the case when A is not isomorphic to the self-adjoint part of a W^*-algebra. By Proposition 2.6.8 its enveloping W^*-algebra $\mathcal{U}$ is σ-finite W^*-factor of type III_λ and A is generated by an involutive *-antiautomorphism α_A on $\mathcal{U}$, that is, $A = \{x \in \mathcal{U}_s : \alpha_A(x) = x\}$. It is known that for any involutive *-antiautomorphism α on $\mathcal{U}$ there are two possible cases: either $mod(\alpha) = 1$, or $mod(\alpha) = \sqrt{\lambda}$ (see Section 2.1).

Hence for a σ-finite JW-factor A of type III_λ we consider separately two cases: $mod(\alpha) = 1$ and $mod(\alpha) = \sqrt{\lambda}$.

Theorem 2.6.9. *Let A be a σ-finite JW-factor of type III_λ, not isomorphic to the self-adjoint part of a W^*-algebra. Then*

1) if $mod(\alpha_A) = 1$, then $D(A) = 2(1 - \lambda^{1/2})/(1 + \lambda^{1/2})$;

2) if $mod(\alpha_A) = \sqrt{\lambda}$, then $D(A) = 2(1 - \lambda^{1/4})/(1 + \lambda^{1/4})$.

Proof. Let $\varphi \in S_0(A)$ and $\overline{\varphi} = (\varphi + \varphi \circ \alpha_A(x))/2$ - α_A-invariant linear extension of φ on $\mathcal{U}$. We denote by $S_0(\mathcal{U})_{\alpha_A}$ the set of all such α_A-invariant normal states on $\mathcal{U}$. Let $Int_R(\mathcal{U})$ denote the set of all inner *-automorphisms of $\Re(A)$, extended onto $\mathcal{U}$ by linearity, and let $[\overline{\varphi}]$ denote the closure in norm of the orbit of $\overline{\varphi}$ in $S_0(\mathcal{U})_{\alpha_A}$. Since each inner *-automorphism of $\Re(A)$ is the product of a finite number of Jordan homomorphisms of the form U_s, where $s \in A$ and $s^2 = \mathbf{I}$ (see [Up, Theorem 1.4]), it follows that the set $S_0(\mathcal{U}_{\alpha_A}) \ / \ Int_R(\mathcal{U})$ of all such classes $[\overline{\varphi}]$, with the metric

$$d\left([\overline{\varphi}], [\overline{\psi}]\right) = \inf\left\{\|\overline{\varphi}' - \overline{\psi}'\| : \quad \overline{\varphi}' \in [\overline{\varphi}],\ \overline{\psi} \in [\overline{\psi}]\right\},$$

is isometrically isomorphic to the metric space $S_0(A) \ / \ Int(A)$ (since $\|\varphi - \psi\| = \|\overline{\varphi} - \overline{\psi}\|$). Using this isomorphism, we are able via the proof of Theorem 2.6.9 to make use of many results which have been shown for states on W^*-algebras.

1). Let $mod(\alpha_A) = 1$. First we show that $D(A) \geq 2(1-\lambda^{1/2})/(1+\lambda^{1/2})$. By Theorem 3.1 of [CHS], if faithful normal functionals φ and ψ on the W^*-algebra $\mathcal{U}$ satisfy the conditions

(a) φ and ψ commute (i.e. their groups of modular *-automorphisms commute) and $a \cdot \varphi \leq \psi \leq b \cdot \varphi$;

(b) $Sp\Delta_\varphi \bigcup (a/b, b/a) = \{1\}$, where a, b are real numbers, $0 < a < b$, then

$$\|\varphi \cdot Adu \; - \; \psi\| \geq \|\varphi - \psi\|$$

for all unitary elements $u \in \mathcal{U}$.

We construct α_A-invariant φ and ψ from $S_0(\mathcal{U})$, satisfying the conditions (a) and (b). By [A 13], there exists an faithful normal state $\varphi_0 \in S_0(\mathcal{U})$ such that

$$\varphi_0 \circ \alpha_A = \varphi_0 \qquad \text{and} \qquad Sp\Delta_{\varphi_0} \subset \{\lambda^n\} \bigcup \{0\}, \quad n \in Z.$$

Then

$$Sp\Delta_{\varphi_0} \bigcap (\lambda, 1/\lambda) = \{1\}.$$

Then $\mathcal{U}_{\varphi_0}$ is a W^*-factor of type II_1, whence $(R_{\varphi_0})_s = \mathcal{U}_{\varphi_0} \bigcap A$ is a purely real JW-factor of type II_1 (see [A 9]). Then there exists a projection $p \in R_{\varphi_0}$, such that $\varphi_0(p) = 1/(1+\sqrt{\lambda})$. Consider the element $m = \sqrt{\lambda} p + \frac{1}{\sqrt{\lambda}} \cdot p^\perp$. It is clear, that $m \in R_{\varphi_0}$. Moreover

$$\sqrt{\lambda} \cdot \mathbf{I} \leq m \leq \frac{1}{\sqrt{\lambda}} \cdot \mathbf{I} \quad \text{and} \quad \varphi_0(m) = 1.$$

We define a functional $\overline{\psi}(x) = \varphi_0(mx)$, $x \in \mathcal{U}(A)$. It is easy to see that $\overline{\psi}$ is a normal state and that φ_0 and $\overline{\psi}$ commute, and $\overline{\psi} \circ \alpha_A = \overline{\psi}$. By Theorem 3.1 from [CHS] $\|\varphi_0 \cdot Adu - \overline{\psi}\| \geq \|\varphi_0 - \overline{\psi}\|$ for all unitary elements u from $\mathcal{U}$ (and hence for each unitary $u \in \Re(A)$).

Put

$$\varphi_1 = \varphi_0|\mathcal{U}_{\varphi_0}, \quad \psi_1 = \overline{\psi}|\mathcal{U}_{\varphi_0}.$$

Then, as in the proof of 3.1 from [CHS],

$$\|\varphi_0 - \overline{\psi}\| \geq \|\varphi_1 - \psi_1\| = \varphi_1(|\mathbf{1} - m|) = 2(1-\lambda^{1/2})/(1+\lambda^{1/2}).$$

Now we'll show that $D(A) \leq 2(1-\lambda^{1/2})/(1+\lambda^{1/2})$. For this purpose we first prove an assertion similar to Lemma 2.3 in [CHS].

Proposition 2.6.10. *Let* $\mathfrak{R}$ *be a* σ*-finite real* W^**-factor of type* III_λ*,* $0<\lambda<1$*;* $\mathcal{U} = \mathfrak{R} + i\mathfrak{R}$ *its enveloping* W^**-factor, and* α *- the involutive *-antiautomorphism of* $\mathcal{U}$*, generating* $\mathfrak{R}$*.*

(1) Let $mod(\alpha) = 1$*;* $T = -2\pi/\ln\lambda$*;* φ_0 *a faithful normal state* $\mathcal{U}$ *such that* $\varphi_0 \circ \alpha = \varphi_0$ *and* $\sigma_T^{\varphi_0} = id$*. Then, for each faithful normal* α*-invariant state* φ *on* $\mathcal{U}$ *there exists* $h \geq 0$*,* $h \in \mathcal{U}_{\varphi_0} \bigcap \mathfrak{R}$ *such that*

(i) $Sph \subset [\lambda a, a]$ *for any* $a > 0$*;*

(ii) there exists a unitary element $u \in \mathfrak{R}$ *such that*

$$\varphi(uxu^*) = \varphi_0(hx) \text{ for all } x \in \mathcal{U}.$$

(2) Let $mod(\alpha) = \sqrt{\lambda}$*;* $T_1 = -4\pi/\ln\lambda$*;* φ_0 *a faithful normal state on* $\mathcal{U}$ *such that* $\varphi_0 \circ \alpha = \varphi_0$ *and* $\sigma_{T_1}^{\varphi_0} = id$*. Then, for each faithful normal* α*-invariant state* φ *on* $\mathcal{U}$ *there exists* $h \geq 0$*,* $h \in \mathcal{U}_{\varphi_0} \bigcap \mathfrak{R}$ *such that*

(i) $Sph \subset [\sqrt{\lambda} a, a]$ *for any* $a > 0$*;*

(ii) there exists a unitary element $u \in \mathfrak{R}$ *such that*

$$\varphi(uxu^*) = \varphi_0(hx) \text{ for all } x \in \mathcal{U}.$$

Proof. We proceed analogously to the proof of Lemma 2.3 in [CHS]. First we prove (1).

Let $v = \left(\dfrac{\partial\varphi}{\partial\varphi_0}\right)_T$ be the derivative of the Connes cocycle at the point $t = T$ (see Section 2.1). Then $\sigma_T^{\varphi}(x) = vxv^*$, whence $\varphi(vxv^*) = \varphi(x)$, and so $v \in \mathcal{U}_\varphi$.

We shall prove that $\alpha(v) = v$.

Lemma 2.6.11. *Let* φ *and* ψ *be faithful normal states (weights) on* $\mathcal{U}$ *such that* $\varphi \circ \alpha = \varphi$, *and* $\psi \circ \alpha = \psi$. *Then*

$$\alpha\left(\left(\frac{\partial\varphi}{\partial\psi}\right)_t \cdot\right) = \left(\frac{\partial\varphi}{\partial\psi}\right)^*_{-t}$$

for all $t \in \mathbf{R}$.

Proof. Let F_2 be the algebra of 2×2 complex matrices, and $\{e_{ij}\}_{i,j=1,2}$ a system of matrix units of F_2. Let θ be a weight on $\mathcal{U} \otimes F_2$, constructed as follows:

$$\theta(x_{11} \otimes e_{11} + x_{22} \otimes e_{22}) = \varphi(x_{11}) + \psi(x_{22}),$$

where $x_{11} \otimes e_{11} + x_{22} \otimes e_{22} \in (\mathcal{U} \otimes F_2)_+$. As we know, the one-parameter family $\{v_t\}_{t\in\mathrm{R}}$ of unitary operators, appearing as derivatives of the Connes cocycle for the weights φ and ψ $\left(\text{i.e. } v_t = \left(\frac{\partial\varphi}{\partial\psi}\right)_t\right)$ is constructed as follows

$$v_t \otimes e_{21} = \sigma^\theta_t(\mathbf{1} \otimes e_{21}).$$

On F_2 there always exists a canonical involutive*-antiautomorphism α_1, conjugate with the system $\{e_{ij}\}_{i,j=1,2}$ and defined as $\alpha_1(e_{ij}) = e_{ji}$, $i,j = 1,2$.

Consider on $\mathcal{U} \otimes F_2$ the involutive *-antiautomorphism $\overline{\alpha}$, obtained by extending the involutive *-antiautomorphism $\alpha \otimes \alpha_1$, acting on the algebraic tensor product $\mathcal{U}\overline{\otimes}F_2$. It is easy to see that $\theta \cdot \overline{\alpha} = \theta$ by construction, since $\varphi \circ \alpha = \varphi$, $\psi \circ \alpha = \psi$ (see [U 1], for example). We consider the element $\alpha(v_t) \otimes e_{21} = \overline{\alpha}(v_t) \otimes e_{21}$. Since $\theta \cdot \overline{\alpha} = \theta$, we have $\overline{\alpha} \circ \sigma^\theta_t = \sigma^\theta_t \circ \overline{\alpha}$ for the group σ^θ of modular *-automorphism θ [G 3]. Then

$$\overline{\alpha}(v_t \otimes e_{21}) = \overline{\alpha}(\sigma^\theta_t(\mathbf{1} \otimes e_{21})) = \sigma^\theta_{-t}(\overline{\alpha}(\mathbf{1} \otimes e_{21})) =$$

$$\sigma^{\theta}_{-t}(\alpha(\mathbf{1}) \otimes \alpha(e_{21})) = \sigma^{\theta}_{-t}(\mathbf{1} \otimes e_{12}) = \sigma^{\theta}_{-t}((\mathbf{1} \otimes e_{21})^*).$$

Since

$$\sigma^{\theta}_{t}(\mathbf{1} \otimes e_{21}) = v_t \otimes e_{21}, \quad \text{for all} \quad t \in \mathbf{R},$$

we have

$$\sigma^{\theta}_{-t}((\mathbf{1} \otimes e_{21})^*) = \sigma^{\theta}_{-t}(\mathbf{1} \otimes e_{21})^* = v_t \otimes e_{12}.$$

We get $\alpha(v_t) \otimes e_{12} = v_t \otimes e_{12}$, whence $\alpha(v_t) = v_t$. ■

From this lemma it follows that $\alpha(v) = v$. Then all spectral projections in the spectral resolution of the unitary element

$$v = \int_0^{2\pi} e^{it} dP_t$$

lie in $\mathfrak{R}(A)$.

We continue the proof of the proposition. By the Riesz representation theorem and the spectral theorem, on the circle $T = \{\gamma \in \mathbf{C} : |\gamma| = 1\}$ there exists a measure μ such that

$$\int_T f(z) d\mu(z) = \varphi_0(f(z))$$

for every Borel function f on T. Let ν be the measure defined by $\nu(B) = \mu(\exp iB)$ for every Borel set $B \in [0, 2\pi]$. In a manner similar to the proof of Lemma 2.3 in [CHS], we consider the function

$$g(s) = \int_{[s,s+2\pi]} \exp(-iT) d\nu(t).$$

This function is decreasing. We denote by $g(+s)$ the right-hand limit of g at the point s, and by $g(-s)$ the left-hand limit of g at the point s. We choose a point $r \in [-2\pi, 0]$ such that $g(+r) \leq 1 \leq g(-r)$, and fix it. There are two possible cases: either r is a point of continuity of g, or g is discontinuous at r. As it is shown in the proof of Lemma 2.3 from [CHS], g is continuous at r if and only if e^{ir} is an eigenvalue of v. We consider both cases separately.

Case 1. Let e^{ir} be not an eigenvalue of v. With the help of v we shall construct the element d in the following way. Let Arg_r :

$T \setminus \{e^{ir}\} \to (r, r+2\pi)$ be a function of the argument with values in $(r, r+2\pi)$. Consider the element $d = Arg_r(v)$. Since all spectral projections v lie in $\mathfrak{R}$, we have $d \in \mathfrak{R}$.

Let $k = \exp(d/T)$. Clearly $k \in \mathfrak{R}$. Then, just as in the proof of Lemma 2.3 from [CHS], $Spd \subset [r, r+2\pi]$, whence

$$Spk \subset [\exp(r/T), \exp(r/T)/\lambda].$$

Since r is a point of continuity for g, we have

$$\varphi(k^{-1}) = \int_T \exp((1/T)Arg_r(z)d\mu(z) = \int_r^{r+2\pi} \exp(-t/T)d\nu(t) = 1$$

Consider the linear functional $\psi(x) = \varphi(k^{-1}x)$, $x \in \mathcal{U}$. Then $\psi \circ \alpha = \psi$, since $k \in \mathfrak{R}$, and ψ is a faithful normal state. Since $k^{iT} = \exp(id) = v$, it follows that $\sigma_T^{\psi}(x) = x$ for all $x \in \mathcal{U}$. Therefore

$$\left(\frac{\partial \psi}{\partial \varphi_0}\right)_T = \left(\frac{\partial \psi}{\partial \varphi}\right)_T \cdot \left(\frac{\partial \varphi}{\partial \varphi_0}\right)_T = k^{-iT} \cdot v = \mathbf{1}.$$

By Lemma 2.6 from [St 4] and by Theorem 3.3(2) from [U 1] it follows that there exists a unitary element $u \in \mathfrak{R}$ such that $\psi \circ Adu = \varphi_0$.

Put $h = u^*ku$. Then

$$\varphi(uxu^*) = \psi(kuxu^*) = \psi(uhxu^*) = \varphi_0(hx)$$

and h and u satisfy the conditions (i) and (ii) of the proposition.

Case 2. Let e^{ir} be an eigenvalue of v. We take p as the projection onto the eigensubspace corresponding to e^{ir}.

We choose, similar to the proof of Lemma 2.3 in [CHS], a number $\beta \in [0,1]$ such that $1 = (1-\beta)g(+r) + \beta g(-r)$. We have $\sigma_T^{\varphi}(x) = vxv^*$ for all $x \in \mathcal{U}$, and $pv = e^{ir}p$. It is easy to see that $p \in \mathcal{U}_{\varphi_0}$. Let us prove that $p \in \mathcal{U}_{\varphi_0} \bigcap \mathfrak{R}$. Assume the opposite $p \notin \mathfrak{R}$. Then $\alpha(p) \neq p$.

For convenience denote $\alpha(p)$ as $\overline{p}$. From the properties of α it follows that

$$\alpha(pv) = v\alpha(p) = e^{ir}\alpha(p) = e^{ir}\overline{p}.$$

Therefore $v\overline{p} = e^{ir}\overline{p}$, whence

$$\overline{p}v^* = e^{-ir}\overline{p}, \qquad pv^*v = e^{-ir}\overline{p}v,$$

i.e. $\overline{p} = e^{-ir}\cdot\overline{p}v$ and $\overline{p}v = e^{ir}\overline{p}$. We have $pv = e^{ir}p$, $\overline{p}v = e^{ir}\overline{p}$. By the maximality of p, it follows that $p = \overline{p}$. Thus $p \in \mathfrak{R}$. Since $p \in \mathcal{U}_{\varphi_0}$, one has $p \in \mathcal{U}_{\varphi_0} \bigcap \mathfrak{R}$.

The restriction of σ_T^φ on $p\mathcal{U}p$ acts trivially. Since $\mathcal{U}$ has type III, $p\mathcal{U}p \simeq \mathcal{U}$. By Proposition 2.1.5 $(p\mathcal{U}p)_{\tilde{\varphi}_0}$ is a W^*-factor of type II_1, where $\tilde{\varphi}_0$ is the restriction of φ_0 to $p\mathcal{U}p$.

By Theorem 2.4 of [A 8], $\left((p\mathcal{U}p)_{\tilde{\varphi}_0} \bigcap \mathfrak{R}\right)_s$ is a purely real JW-factor of type II_1. Then there exists a projection $p' \leq p$, $p' \in (p\mathcal{U}p)_{\tilde{\varphi}_0} \bigcap \mathfrak{R}$, such that $\varphi(p') = \beta\varphi(p)$. We define d and k as follows:

$$d = Arg_r\left(vp^\perp + rp' + (r+2\pi)(p-p')\right); \quad k = \exp(d/T).$$

It is easy to see that $d \in \mathfrak{R}$, $k \in \mathfrak{R}$. Now the proof proceeds just as in case I. The constructed h and u are in $\mathfrak{R}$, and $Sph \subset [\lambda a, a]$.

Now we shall prove the assertion (2) of the proposition. Unlike to the case (1), since $mod(\alpha) = \sqrt{\lambda}$, on $\mathcal{U}$ there exists no faithful normal state φ_0 such that $\varphi_0 \circ \alpha = \varphi_0$, $\sigma_T^{\varphi_0} = id$, where $T = -2\pi/\ln\lambda$. Let $\overline{\varphi}_0$ be a faithful normal strictly semifinite α-invariant weight such that $\sigma_{T_1}^{\overline{\varphi}_0} = id$, where $T_1 = 2T = -4\pi/\ln\lambda$ (the existence of such weight was proved in Proposition 2.1.4 (b)). If we take any projection $e \in \mathcal{U}_{\overline{\varphi}_0} \bigcap \mathfrak{R}$ such that $\overline{\varphi}_0(e) < +\infty$, and consider the isomorphism $\sigma : e\mathcal{U}e \to \mathcal{U}$ constructed in Theorem 3.2 of [U 1] such that $\sigma \circ \alpha = \alpha \circ \sigma$, we obtain a bounded linear functional $\varphi_0' = \overline{\varphi}_0|_{e\mathcal{U}e} \circ \sigma^{-1}$ on $\mathcal{U}$ with the properties

$$\varphi_0' \circ \alpha = \varphi_0', \quad \sigma_{T_1}^{\varphi_0'} = id.$$

Normalizing it, we get a faithful normal α-invariant state φ_0 such that

$$\sigma_{T_1}^{\varphi_0} = id \text{ and } Sp\Delta_{\varphi_0} \subset \{\lambda^{n/2}\} \bigcup \{0\},\ n \in Z.$$

Now the proof is carried over similar to (1) without change, except for the following point. Since $T_1 = 2T, Spd \subset [r, r + 2\pi]$ and $k = \exp(d/T_1)$ it follows that

$$Spk \subset [\exp(r/T_1),\ \exp((r + 2\pi)/T_1)],$$

that is,

$$Spk \subset [\exp(r/T_1),\ (1/\sqrt{\lambda}\exp(r/T_1)]. \quad \blacksquare$$

Now we shall continue the proof of Theorem 2.6.9 and prove the inequality $D(A) \leq 2(1 - \lambda^{1/2}/(1 + \lambda^{1/2})$. Let φ_0 be a faithful normal state of $\mathcal{U}$ such that $\varphi_0 \circ \alpha_A = \varphi_0$ and $\sigma_T^{\varphi_0} = id$, for $T = -2\pi/\ln\lambda$, and let φ and ψ be arbitrary faithful normal states of $\mathcal{U}$. By 2.6.10 (1) there exist $\varphi' \in [\varphi]$ and $\psi' \in [\psi]$ such that

$$\varphi'(x) = \varphi_0(hx), \quad \psi'(x) = \varphi_0(kx) \text{ for all } x \in \mathcal{U},$$

where h and k are positive elements from $\mathcal{U}_{\varphi_0} \bigcap \Re$ and $\lambda a\mathbf{1} \leq h \leq a\mathbf{1}$, $\lambda b\mathbf{1} \leq k \leq b\mathbf{1}$ for some $a > 0, b > 0$. Consider any number $\delta > 0$. By the spectral theorem there exist two families of pairwise orthogonal projections

$$\{p_i\}_{i=1}^n \subset \mathcal{U}_{\varphi_0} \bigcap \Re(A), \quad \{q_i\}_{i=1}^n \subset \mathcal{U}_{\varphi_0} \bigcap \Re(A),$$

such that $\varphi_0(p_i) = 1/n$, and $\varphi_0(q_i) = 1/n$, $i \in \overline{1, n}$ and

$$\left\| h - \sum_{i=1}^n \beta_i p_i \right\|_1 < \delta, \qquad \left\| k - \sum_{i=1}^n \gamma_i q_i \right\|_1 < \delta$$

for some $\{\beta_i\}_{i=1}^n$ and $\{\gamma_i\}_{i=1}^n$ from $\mathbf{R}$, where $\beta_1 \geq \beta_2 \geq \ldots \geq \beta_n = \lambda\beta_1$, and $\gamma_1 \geq \gamma_2 \geq \ldots \geq \gamma_n = \lambda\gamma_1$, with $\sum \beta_i = \sum \gamma_i = n$, while the norm $\|\ \|_1$ is defined as $\|x\|_1 = \varphi_0(|x|)$, $x \in \mathcal{U}$.

Without lost of generality we may put

$$h = \sum_{i=1}^{n} \beta_i p_i, \quad \text{and} \quad k = \sum_{i=1}^{n} \gamma_i q_i.$$

Since $\mathcal{U}_{\varphi_0}$ is a factor of type II_1, it is easy to see that $\mathcal{U}_{\varphi_0} \bigcap A$ is a purely real JW-factor of type II_1. In a manner similar to Lemma 2.6.6 there exists a unitary element $u \in \mathcal{U}_{\varphi_0} \bigcap \Re(A)$ such that $uq_iu^* = p_i$, $i = \overline{1,n}$. Then $uku^* = \sum_{i=1}^{n} \gamma_i p_i$.

Consider the state ψ'' of the form

$$\psi''(x) = \varphi_0(uku^*x) = \varphi_0(ku^*xu).$$

It is clear that $\psi'' \in [\psi]$ and that $\psi'' \circ \alpha_A = \psi''$.

Let f and g be functions on $[\lambda, 1]$, defined by

$$f(t) = \frac{1}{1-\lambda} \sum_{i=1}^{n} \beta_i \, \chi_{I_i}(t), \qquad g(t) = \frac{1}{1-\lambda} \sum_{i=1}^{n} \gamma_i \, \chi_{I_i}(t),$$

where

$$I_i = \begin{cases} [\lambda + (i-1)\frac{(1-\lambda)}{n}, \ \lambda + i\frac{(1-\lambda)}{n}], & \text{if } \ i \in \overline{1, n-1} \\ [\lambda + (n-1)\frac{(1-\lambda)}{n}, \ 1], & \text{if } \ i = n \end{cases}$$

Then, by Corollary 2.2 of [CHS],

$$\|\varphi' - \psi'\| = \|h - uku^*\|_1 = \sum_{i=1}^{n} |\beta_i - \gamma_i| \varphi_0(p_i) = \int_{\lambda}^{1} |f(t) - g(t)| dt.$$

By Lemma 2.1 of [CHS],

$$\int_{\lambda}^{1} |f(t) - g(t)| dt \le 2(1 - \lambda^{1/2})/(1 + \lambda^{1/2}).$$

Thus we have $D(A) = 2(1-\lambda^{1/2})/(1+\lambda^{1/2})$ in the case when $mod(\alpha_A) = 1$.

Now we shall prove the assertion (2) of the theorem. Suppose that $mod(\alpha_A) = \sqrt{\lambda}$. We'll prove first that $D(A) \geq 2(1-\lambda^{1/4})/(1+\lambda^{1/4})$. As we know there exists a faithful normal state φ_0 on $\mathcal{U}$ such that

$$\varphi_0 \circ \alpha_A = \varphi_0, \qquad Sp\Delta_{\varphi_0} = \{\lambda^{n/2}\} \bigcup \{0\},\ n \in Z.$$

Then $Sp\Delta_{\varphi_0} \bigcap (\sqrt{\lambda}, 1/\sqrt{\lambda} = \{1\}$. Repeating entirely the arguments in the proof of assertion (1) (the inequalities $D(A) \geq 2(1-\lambda^{1/2})/(1+\lambda^{1/2})$), with the only difference that everywhere instead of λ we put $\sqrt{\lambda}$, we obtain

$$D(A) \geq 2(1-\lambda^{1/4})/(1+\lambda^{1/4}).$$

Now let us show that

$$D(A) \leq 2(1-\lambda^{1/4})/(1+\lambda^{1/4}).$$

We proceed analogously the proof of assertion (1). Let φ_0 be a faithful normal state $\mathcal{U}$ such that $\varphi_0 \circ \alpha_A = \varphi_0$, and $\sigma_{T_1}^{\varphi_0} = id$, where $T_1 = -4\pi/\ln\lambda$. By Proposition 2.6.6 (2), if ψ and φ are faithful normal states of $\mathcal{U}$ and $\varphi \circ \alpha_A = \varphi$ and $\psi \circ \alpha_A = \psi$, there exist $\varphi' \in [\varphi]$ and $\psi' \in [\psi]$ such that

$$\varphi'(x) = \varphi_0(hx), \quad \psi'(x) = \varphi_0(kx) \quad \text{for all } x \in \mathcal{U},$$

where h and k are positive elements from $\mathcal{U}_{\varphi_0} \bigcap \Re(A)$ and $\sqrt{\lambda} a\mathbf{1} \leq h \leq a\mathbf{1}$, $\sqrt{\lambda} b\mathbf{1} \leq k \leq b\mathbf{1}$ for some $a > 0, b > 0$. For any $\delta > 0$ there exist families $\{p_i\}_{i=1}^n \subset \mathcal{U}_{\varphi_0} \bigcap \Re(A)$, and $\{q_i\}_{i=1}^n \subset \mathcal{U}_{\varphi_0} \bigcap \Re(A)$, of pairwise orthogonal projections from $\mathcal{U}_{\varphi_0} \bigcap A$ such that $\varphi_0(p_i) = 1/n$, $\varphi_0(q_i) = 1/n$, $i \in \overline{1,n}$ and

$$\left\| h - \sum_{i=1}^n \beta_i p_i \right\|_1 < \delta, \qquad \left\| k - \sum_{i=1}^n \gamma_i q_i \right\|_1 < \delta$$

for appropriate constants $\{\beta_i\}_{i=1}^n$ and $\{\gamma_i\}_{i=1}^n$ with the properties

$$\beta_1 \geq \beta_2 \geq ... \geq \beta_n = \sqrt{\lambda}\beta_1, \quad \gamma_1 \geq \gamma_2 \geq ... \geq \gamma_n = \sqrt{\lambda}\gamma_1,$$

$$\sum \beta_i = \sum \gamma_i = n.$$

where $\|x\|_1 = \varphi_0(|x|)$ for all $x \in \mathcal{U}$.

By Proposition 2.1.5 $\mathcal{U}_{\varphi_0}$ is the direct sum of two W^*-factors of type II_1 and $\mathcal{U}_{\varphi_0} \bigcap A$ is a purely JW-factor of type II_1; as in Lemma 2.6.6, there exists a unitary element $u \in \mathcal{U}_{\varphi_0} \bigcap \Re(A)$ such that $uq_iu^* = p_i$ for all $i \in \overline{1,n}$. Then

$$uku^* = \sum_{i=1}^n \gamma_i p_i.$$

As in case (1), we consider a state ψ'' of the form $\psi''(x) = \varphi_0(uku^*x) = \varphi_0(ku^*xu)$. It is clear that $\psi'' \in [\psi]$, $\psi'' \circ \alpha_A = \psi''$.

Consider the functions f and g on the interval $[\sqrt{\lambda}\ , 1]$, defined as

$$f(t) = \left(\frac{1}{(1-\sqrt{\lambda})}\right) \sum_{i=1}^n \beta_i\ \chi_{I_i}(t),$$

$$g(t) = \left(\frac{1}{(1-\sqrt{\lambda})}\right) \sum_{i=1}^n \gamma_i\ \chi_{I_i}(t),$$

where

$$I_i = \begin{cases} [\sqrt{\lambda} + (i-1)(1-\sqrt{\lambda})/n, \sqrt{\lambda} + i(1-\sqrt{\lambda})/n], & \text{if} \quad i \in \overline{1, n-1}, \\ [\sqrt{\lambda} + (n-1)(1-\sqrt{\lambda})/n,\ 1], & \text{if} \quad i = n. \end{cases}$$

We estimate $\|\varphi' - \psi''\|$, using Lemma 2.1 of [CHS]:

$$\|\varphi' - \psi''\| = \|h - uku^*\|_1 = \sum_{i=1}^n |\beta_i - \gamma_i|\varphi_0(p_i) =$$

$$= \int_{\sqrt{\lambda}}^{1} |f(t) - g(t)| dt \ \leq \ 2(1 - \lambda^{1/4})/(1 + \lambda^{1/4}).$$

Thus $D(A) = 2(1 - \lambda^{1/4})/(1 + \lambda^{1/4})$, if $mod(\alpha_A) = \sqrt{\lambda}$. ■

COMMENTS TO CHAPTER 2

2.1. The discrete decomposition of type III_λ factors, $0 \leq \lambda < 1$ was obtained by Connes in his fundamental paper [C 1]. In fact this reduces the study of type III factors to the investigation of semifinite algebras and their *-automorphisms. By the definition a real factor has type III_λ if its enveloping (complex) factor has the type III_λ, $0 \leq \lambda \leq 1$ [U 3]. For real factors $\mathfrak{R}$ of type III_λ, $0 < \lambda < 1$, the discrete decomposition was proved independently by Usmanov and Stacey for the case when $mod(\alpha_{\mathfrak{R}}) = 1$, and in the case of $mod(\alpha_{\mathfrak{R}}) = \sqrt{\lambda}$ this was done by Stacey [St 4], [U 1], [U 3], [U 4]. Our exposition is close to the paper [St 4].

2.2. The results of this Section are due to Usmanov [U 2],[U 4]. For real factors of type III_0 with $mod(\alpha_{\mathfrak{R}}) \neq 1$ the discrete decomposition is still open; perhaps it will be similar to the construction in Theorem 2.1.9 (b).

2.3. The classification of periodic *-automorphisms of hyperfinite factors was given by Connes in his papers [C 5], [C 3]. The classification of involutive (i.e. of period 2) *-antiautomorphisms of hyperfinite factors was obtained by Giordano in [G 2], [G 3], [G 4], which enabled us to classify hyperfinite JW-factors and real factors, generated by involutive *-antiautomorphisms. In its turn this has motivated the study of periodic *-antiautomorphisms of hyperfinite factors and periodic *-automorphisms of real hyperfinite factors. In this Section the results of Rakhimov and Usmanov are exposed [R 1],[R 2], [RU 3], [RU 4]. Theorem 2.3.21 is a real analogue of Connes result. Theorem 2.3.17 is a generalization of a result Giordano: it was also proved independently by Giordano [G 5]. The example following Theorem 2.3.23, exposing the essential difference between *-antiautomorphisms of real W^*-algebras and *-antiautomorphisms of complex W^*-algebras, was constructed by Rakhimov [R 2], [RU 4].

2.4. In papers of Connes [C 2], [C 3] outer conjugacy classes of *-automorphisms were described. In this Section we give a description of outer conjugacy classes of *-antiautomorphisms of hyperfinite factors and *-automorphisms of real hyperfinite factors following the results of Rakhimov and Usmanov [RU 6]. The technique of proofs are similar to Connes [C 2]. Lemma 2.4.4 is a real analogue of the non-commutative Rohlin's lemma obtained by Connes in [C 2]. Corollary 2.4.15 presents also an alternative proof of Theorem 1.7.3 (iii).

2.5. The equivalence of injectivity, amenability, semifiniteness and hyperfiniteness for W^*-algebras was proved by Connes [C 4]. Real versions of hyperfinite W^*-algebras were considered by Stormer, who proved the equivalence of injectivity and hyperfiniteness of real II_1 factors [S 5]. The equivalence of injectivity, amenability, semidiscreteness and hyperfiniteness for real W^*-algebras $\mathfrak{R}$ satisfying the condition $mod(\alpha_{\mathfrak{R}}) = 1$, was proved by Boykabilov [Boy 1], [Boy 2]. Theorem 2.5.15 exposing that a real injective factor may be not hyperfinite was obtained by Rakhimov [R 6].

2.6. Diameters of the state spaces for W^*-algebras were investigated in [CHS],[CS]. For JW-factors of all types except III_0 and III_1 they were calculated by Ayupov and Usmanov [A 11], [AU 1], [AU 2]. The case of type III_0 JW-factors remains open.

CHAPTER 3

LIE STRUCTURE IN OPERATOR ALGEBRAS

Let M be a von Neumann algebras, α - involutive *-anti-automorhism on M. Consider $M^{\alpha}(\pm 1) = \{x \in M : \alpha(x) = \pm x\}$ the spectral subspaces of α. It follows that $M^{\alpha}(+1)$ is a Jordan algebra with respect to the symmetrized product $x \cdot y = \frac{1}{2}(xy + yx)$ and $M^{\alpha}(-1)$ is a Lie algebra with pespect to the brackets $[x, y] = xy - yx$. Robinson and Stormer [RS] have initiated the classification of pairs of Jordan and Lie algebras which can occur in this manner by examining the case $M = B(H)$ the algebra of all bounded linear operators on a complex Hilbert space H, i.e. when M is a type I factor. In this chapter we shall continue the study of Lie algebras $M^{\alpha}(-1)$ for arbitrary factors M.

The main problem considered in this chapter is the proof of the following conjecture: Lie algebras $M^{\alpha}(-1)$ and $M^{\beta}(-1)$ are isomorphic if, and only if, the anti-automorphisms α and β are conjugate, i.e. $\alpha = \theta^{-1}\beta\theta$ for a *-automorphism θ of M.

The proof of this conjecture is reduced to study a Lie isomorphism between $\Re_k$ and S_k - the Lie algebras of skew elements of real factors $\Re$ and S, and the problem of their extension to an associative isomorphism between $\mathfrak{R}$ and S. This problem turns out to be a special case of a classical algebraic problem: letting $\mathfrak{R}$ and S be simple or prime rings with involutions, can every Lie isomorphism of $[\mathfrak{R}_k, \mathfrak{R}_k]$ onto $[S_k, S_k]$ be lifted to an associative isomorphism of $\mathfrak{R}$ onto S. In order to solve the last problems in Sections 3.2 and 3.3 we show that every elements in $\mathfrak{R}$ can be represented by means of elements of the Lie algebra $[\mathfrak{R}_k, \mathfrak{R}_k]$.

In Section 3.1 we give some preliminaries from the theory of prime algebras with involution and prove that any real factor is a *-closed prime algebra over $\mathbf{R}$ with involution of the first kind.

In Section 3.3 we prove the following mean result:
if $\mathfrak{R}$ and S are real factors not of type I_1 and I_2 then every Lie isomorphism between $[\mathfrak{R}_k, \mathfrak{R}_k]$ and $[S_k, S_k]$ can be uniquely lifted to an associative *-isomorphism of $\mathfrak{R}$ onto S (see Theorem 3.3.10).

In Section 3.4 we define Lie algebra $M^\alpha(-1)$, establish their connection with Lie algebras of skew-adjoint elements of real von Neumann algebra, and consider some examples. Theorem 3.3.10 implies the theorem of isomorphism for Lie algebras of the form $M^\alpha(-1)$.

3.1. Theorem on isomorphism of prime Lie rings

Let A be an associative algebra over a field F. Recall that A is said to be *semiprime*, if $aAa = \{0\}$ implies $a = 0$; A is said to be *prime*, if $aAb = \{0\}$ implies $a = 0$ or $b = 0$. A is said to be *simple*, if it dosn't contain nonzero (two-sided) ideals. Following [CR] we shall introduce the notion of the extended centroid of A. *Partially defined centralizer* (p.d.c.) on A is a linear operator f from A into A such that $dom(f)$ (domain of f) is a (two-sided) ideal in A and $f(ab) = af(b)$, $f(ba) = f(b)a$ for all $a \in A, b \in dom(f)$. *Essentially defined centralizer* (e.d.c.) on A is a p.d.c. with $dom(f)$ being an essential ideal in A, i.e. $dom(f) \bigcap U \neq \{0\}$ for any nonzero ideal U in A. A partially defined centralizer f is said to be *maximal* if there is no p.d.c. g on A, $g \neq f$, extending f, i.e. with $dom(f) \subset dom(g)$ and $g(x) = f(x)$ for all $x \in dom(f)$.

Proposition 3.1.1[CR, Proposition 1]. *Given any e.d.c. f on a semiprime algebra A there exists a unique maximal e.d.c. $\tilde{f}$ on A, extending f.*

Now the main result of [CR, Theorem 1] on extended centroids can be stated as follows.

Proposition 3.1.2. *Let A be a semiprime algebra. The set $C(A)$ of all maximal e.d.c. on A equipped with usual algebraic operations forms a von Neumann regular commutative ring called the extended centroid of A. Moreover A is prime if and only if $C(A)$ is a field.*

It should be mentioned that a particular case of maximal e.d.c. are totally defined centralizers, which form the set $B(A)$ - *centroid* of A. It is clear that $B(A)$ is a subring of $C(A)$.

If A is prime algebra with involution $x \to x^*$, then one can consider the induced involution on the field $C = C(A)$. The set C_* of all symmetric elements of C is also a field which is called the *-extended centroid of A.

Definition. A prime algebra A with involution over a field F is said to be **-closed* if $C_* = F$, i.e. F is already the *-extended centroid of A.

We say the involution on A is of the *first kind* if it induces the identity involution on C, i.e. $C_* = C$. Otherwise it is of the *second kind* which means that C contains a nonzero skew element γ, in this case $C = C_* + \gamma C_*$.

It is easy to show that any W^*-algebra (real or complex) R is semiprime, and R is prime if, and only if R is factor.

Proposition 3.1.3. *A real factor $\Re$ is simple if, and only if $\mathcal{U} = \Re + i\Re$ is a simple factor.*

Proof. Let $\mathcal{U}$ be simple and suppose that J is a proper ideal in $\Re$. Then $J + iJ$ is a proper ideal in $\mathcal{U}$ and hence $J + iJ = \{0\}$. Therefore $J = \{0\}$, and $\Re$ is simple.

Conversely, let $\Re$ be a simple real factor and suppose that J is a nonzero proper ideal in $\mathcal{U}$. Since $\mathcal{U}$ is a factor J is weakly dense in $\mathcal{U}$. Consider $J \cap \Re$ which is an ideal in $\Re$. Since $\Re$ is simple there are only two possibilities:

(1) $\quad J \cap \Re = \Re$,

or

(2) $\quad J \cap \Re = \{0\}$;

Let us show that each of them leads to a contradiction.

If $J \cap \Re = \Re$, i.e. $J \supset \Re$, then $J \supset \Re + i\Re = \mathcal{U}$, because J is a complex subspace of $\mathcal{U}$. This is impossible since J is proper.

Suppose that $J \cap \Re = \{0\}$. First let us prove that J contains a net of projections increasing to $\mathbf{1}$. Since the set of projections in the ideal J is a sublattice in the projection lattice of $\mathcal{U}$ [T 1, p.309, Exercise 5(b)], it is sufficient to show that any nonzero projection $f \in \mathcal{U}$ majorizes a nonzero projection $e \in J$. Passing, if necessary, to the subalgebra $f\mathcal{U}f$ and its weakly dense ideal fJf, we may assume without loss of generality that $f = \mathbf{1}$. So we have to prove that J contains a nonzero projection. Let $x \in J, x \neq 0$. Then x^*x is a nonzero positive element in J and, by the spectral theorem, there exists a nonzero projection e, commuting with x^*x, and a real $\lambda > 0$ such that $e \leq \lambda x^*x$. Since J is an ideal $e \in J$. So there exists a net $\{e_\alpha\}$ in J increasing to $\mathbf{1}$.

Now let e be an arbitrary projection in $J \subset \mathcal{U} = \Re + i\Re$, i.e. $e = a + ib$, $a, b \in \Re$. From $\Re \bigcap i\Re = \{0\}$ it follows that $a = a^*$, $b^* = -b$, and moreover $a \geq 0$ (Corollary 1.1.4). Further, from

$$e = e^2 = a^2 - b^2 + i(ab + ba)$$

we have

$$a = a^2 - b^2, \quad \text{i.e.} \quad b^2 = a^2 - a.$$

Since $a = e - ib$, i.e. $a^2 = e - ieb - ibe - b^2$, it follows that

$$a^2 + b^2 = e - ibe - ieb \in J.$$

But $b^2 = a^2 - a$, so $2a^2 - a \in J$. At the same time $2a^2 - a \in \mathfrak{R}$, i.e. $2a^2 - a \in J \bigcap \mathfrak{R} = \{0\}$. Therefore $a = 2a^2$. By the spectral theorem the last equality means that $a = \frac{1}{2}f$, where f is a projection in $\mathfrak{R}$. It is clear that $fa = a$ and thus $fb^2 = f(a^2 - a) = a^2 - a = b^2$, i.e. $(\mathbf{I} - f)b^2 = 0$. Therefore

$$((\mathbf{I} - f)b)((\mathbf{I} - f)b)^* = -(\mathbf{I} - f)b^2(\mathbf{I} - f) = 0,$$

i.e.

$$(\mathbf{I} - f)b = 0 \quad \text{and} \quad fb = b.$$

Hence

$$fe = f(a + ib) = a + ib = e,$$

i.e. $f \geq e$. This means that any projection $e \in J$ has the form

$$e = \frac{1}{2}f + ib,$$

where $f \in \mathfrak{R}$ is a projection, $f \geq e$, and $b \in \mathfrak{R}$.

Now let $\{e_\alpha\}$ be a net of projections from J increasing to $\mathbf{I}$, the existence of which was proved above. Then

$$e_\alpha = \frac{1}{2}f_\alpha + ib_\alpha,$$

where f_α are projections from $\mathfrak{R}$, $f_\alpha \geq e_\alpha$, $b_\alpha \in \mathfrak{R}$. If $e_\alpha \geq e_\beta$, then

$$0 \leq e_\alpha - e_\beta = \frac{1}{2}(f_\alpha - f_\beta) + i(b_\alpha - b_\beta)$$

and by Corollary 1.1.4 $f_\alpha - f_\beta \geq 0$, i.e. $\{f_\alpha\}$ is also increasing. Moreover $f_\alpha \geq e_\alpha$, so the net $\{f_\alpha\}$ is increasing to $\mathbf{I}$. In particular, the nets $\{e_\alpha\}$ and $\{f_\alpha\}$ weakly converge to $\mathbf{I}$, and therefore the net

$$b_\alpha = i(\frac{1}{2}f_\alpha - e_\alpha) \in \mathfrak{R}$$

weakly converges to $i(\frac{1}{2}\mathbf{1}-\mathbf{1}) = -i\frac{1}{2}\mathbf{1} \in i\Re$. But $\Re$ is weakly closed, so $-i\frac{1}{2}\mathbf{1} \in \Re \bigcap i\Re = \{0\}$. This contradiction shows that the case $J \cap \Re = \{0\}$ is also impossible. This means that $\mathcal{U}$ doesn't contain nonzero proper ideals, i.e. $\mathcal{U}$ is simple. ■

Remark 3.1.4. The last proposition 3.1.3 is not valid for general real W^*-algebras (not factors).

Indeed, let

$$\Re = \{(\lambda, \overline{\lambda}),\ \lambda \in \mathbf{C}\}.$$

Then $\Re$ is a real W^*-algebra isomorphic to the field $\mathbf{C}$. In particular $\Re$ is simple. But

$$\mathcal{U} = \Re + i\Re = \{(\lambda, \mu),\ \lambda, \mu \in \mathbf{C}\} = \mathbf{C} + \mathbf{C},$$

is not simple (moreover it is not a factor) because $\mathcal{U}$ has nonzero proper ideals

$$\{(0, \mu), \mu \in \mathbf{C}\} \quad \text{and} \quad \{(\lambda, 0), \lambda \in \mathbf{C}\}.$$

Theorem 3.1.5. *A real factor $\Re$ is simple if it is either finite or σ-finite type III.*

Proof. From corollary 1.2.4 and theorem 1.3.2 we know that a real factor $\Re$ is finite (respectively of type III) if, and only if, $\mathcal{U}$ is finite (respectively of type III). Further, a W^*-algebra $\mathcal{U}$ is simple if it either finite factor or σ-finite type III factor (see [T 1, 309 p., Exercise 5(f)]). Therefore the proof follows from Propositions 3.1.3 and 1.8.8. ■

It is known [M] that the extended centroid of a prime C^*-algebra (in particular of a complex factor) coincides with the field $\mathbf{C}$ of complex numbers. Using this result we shall obtain a similar one for real factors.

Proposition 3.1.6. *For any real factor $\mathfrak{R}$ the extended centroid $C(\mathfrak{R})$ coincides with the real numbers* $\mathbf{R}$. *In particular $\mathfrak{R}$ is a *-closed prime algebra over* $\mathbf{R}$ *with involution of the first kind.*

Proof. Let $\mathcal{U} = \mathfrak{R} + i\mathfrak{R}$ be the complex envelope of $\mathfrak{R}$. Then $\mathcal{U}$ is a complex factor and therefore $C(\mathcal{U}) = \mathbf{C}$. Let us show that $C(\mathfrak{R})$ can be embedded as a subring of $C(\mathcal{U})$. Let $f \in C(\mathfrak{R})$, i.e. $f : U \to \mathfrak{R}$ is a maximal e.d.c. on $\mathfrak{R}$, where $U = dom(f)$ is an essential ideal of $\mathfrak{R}$. Then $U + iU$ is an essential ideal of $\mathcal{U}$. Indeed, given any nonzero ideal J in $\mathcal{U}$ the intersection $J \cap \mathfrak{R}$ is a nonzero ideal in $\mathfrak{R}$ (see the proof of Proposition 3.1.3). Therefore $U \cap (J \cap \mathfrak{R}) \neq \{0\}$ since U is an essential ideal of $\mathfrak{R}$. Thus

$$(U + iU)\bigcap J \supset U\bigcap J \supset U\bigcap(J\bigcap\mathfrak{R}) \neq \{0\},$$

i.e. $U + iU$ is an essential ideal in $\mathcal{U}$. Now e.d.c. $f \in U \to \mathfrak{R}$ can be extended by the complex linearity to the operator

$$f_0 : U + iU \to \mathfrak{R} + i\mathfrak{R} = \mathcal{U},$$

which is e.d.c. on $\mathcal{U}$. By Proposition 3.1.1 f_0 can be extended in a unique way to a maximal e.d.c. $\tilde{f}_0$ on $\mathcal{U}$. Thus the map $f \to \tilde{f}_0$ is the needed embedding of $C(\mathfrak{R})$ into $C(\mathcal{U})$. Since $C(\mathcal{U}) = \mathbf{C}$, each element $g \in C(\mathcal{U})$ has the form $g(x) = \lambda x$, $x \in \mathcal{U}$, for some $\lambda \in \mathbf{C}$. In particular $\tilde{f}_0(x) = \lambda_0 x$ for an appropriate $\lambda_0 \in \mathbf{C}$. But $\tilde{f}_0(\mathfrak{R}) \subset \mathfrak{R}$, i.e. $\lambda_0 x \in \mathfrak{R}$ for all $x \in \mathfrak{R}$. Since $\mathfrak{R}\bigcap i\mathfrak{R} = \{0\}$, we necessarily have that $\lambda_0 \in \mathbf{R}$, i.e. any $f \in C(\mathfrak{R})$ is of the form $f(x) = \lambda x$, $x \in \mathfrak{R}$, for some $\lambda \in \mathbf{R}$. Therefore

$$C(\mathfrak{R})_* = C(\mathfrak{R}) = \mathbf{R}\,.$$

In particular $\mathfrak{R}$ is a *-closed prime algebra over $\mathbf{R}$ with involution of the first kind. ■

Now let A be an arbitrary algebra over a field F. It is said to be *PI-algebra* over F if it satisfies a polynomial identity, i.e. there exists a nonzero polynomial $f = f(x_1, ..., x_n)$ over F such that $f(a_1, ..., a_n) = 0$

for all $a_1, ..., a_n \in A$.

Proposition 3.1.7. *A real factor $\mathfrak{R}$ is a PI-algebra if, and only if $\mathfrak{R}$ is of type I_n $(n < \infty)$.*

Proof. Since type I_n factors are exactly finite dimensional matrix algebras, the "if" part is clear (see [He 3, §6]). So assume that $\mathfrak{R}$ is not of type I_n $(n < \infty)$, and let us show that $\mathfrak{R}$ is not a PI-algebra. Suppose the opposite, i.e. let $\mathfrak{R}$ satisfy a polynomial identity

$$f(a_1, ..., a_n) = 0,$$

for a polynomial f of degree ℓ, and all $a_1, a_2, ..., a_n \in \mathfrak{R}$. Since $\mathfrak{R}$ is a factor and not type I_n $(n < \infty)$, given any ℓ there exists a *-subalgebra A in $\mathfrak{R}$ isomorphic to the $\ell \times \ell$ matrix algebra (cf. [T 1, Chap. IV, Proposition 1.8]). But the $\ell \times \ell$ matrix algebra A cannot satisfy any polynomial identity of degree less than 2ℓ [He 3, Lemma 6.1.3] - a contradiction. ■

Given an algebra A with involution, denote by A_k and A_s respectively the subspaces of skew-adjoint and self-adjoint elements in A, i.e.

$$A_k = \{x \in A : x^* = -x\},$$

and

$$A_s = \{x \in A : x^* = x\}.$$

We denote by $[A_k, A_k]$ the Lie ideal of "derived" skew elements, the linear span of all commutators $[a, b] = ab - ba$, $a, b \in A_k$. Now the main result of Martindale [Ma 2] or Rozen [Roz] can be reformulated for simplicity as follows.

Theorem 3.1.8. *Let $\mathfrak{R}$ and S be *-closed prime algebras with involution of the first kind over a field F of characteristic $\neq 2$. Assume that*

(a) neither $\mathfrak{R}$ nor S is PI-algebra over F;

(b) $\mathfrak{R}$ contains three nonzero orthogonal symmetric idempotents e_1, e_2, e_3 such that $e_1 + e_2 + e_3 = \mathbf{1}$.

Set

$\Re^{ij} = e_i \Re e_j, \quad i,j = 1,2,3, \ i \neq j;$

$\Re^i = e_i \Re e_i, \quad i,j = 1,2,3;$

$\mathcal{U}^i = [\Re_k, \Re_k] \bigcap \Re^i, \ i = 1,2,3,$

and suppose that

(c) $e_i \in \overline{\mathcal{U}}^i$, $i = 1,2,3$, *where* $\overline{\mathcal{U}}^i$ *denotes the *-subalgebra in* $\Re^i$ *generated by* $\mathcal{U}^i$.

Let φ *be a Lie isomorphism between* $[\Re_k, \Re_k]$ *and* $[S_k, S_k]$.

*Then there exists a unique 'related' associative *-isomorphism* φ_0 *of* W *into* S, *where* W *is the associative subalgebra of* $\Re$ *generated by* $\{\mathcal{U}^i, \Re^{ij} \mid i,j = 1,2,3, \ i \neq j\}$.

3.2. Symmetric and skew-symmetric operators on real and quaternionian Hilbert spaces.

Let H will denote a real Hilbert space of dimension n $(n = 1, 2, \dots \infty)$, $B(H)$ - the algebra of all bounded linear operators on H. Then $B(H)$ is a real factor and therefore a prime involutive algebra, which is simple if and only if $n < \infty$. W.Baxter [Bax] proved that if $\Re$ is a simple ring of characteristic not 2 with an involution, which is more than 16-dimentional over its center, then every symmetric elements in $\Re$ can be written as a sum of squares of skew-symmetric elements of $\Re$ with coefficients $\pm$ 1. In the following theorem this result of Baxter is extended to the non simple ring $\Re = B(H)$.

Theorem 3.2.1. *Let* $dim(H) = n > 2$, *and* A *be a symmetric operator in* $B(H)$. *Then there are twelve skew-symmetric operators* $K_1, K_2, \dots, K_{12} \in B(H)$ *such that*

$$A = \sum_{i=1}^{12} \varepsilon_i K_i^2, \quad \text{where } \varepsilon_i = \pm 1, \ i = 1, \dots, 12.$$

In order to prove theorem 3.2.1 we need some preliminary results. The first two lemmas reduce the general case to the case when $H = L_2(X,m)$ - the real L_2-space, and the symmetric operator A coincides with M_a - operator of multiplication by a bounded real function a on X.

Lemma 3.2.2. *Given any symmetric bounded linear operator A on a real Hilbert space H, there exists a measure space (X,m), a bounded real function a on X and an orthogonal operator $O : H \to L_2(X,m)$ such that*

$$A = O^{-1} M_a O;$$

moreover X may be identified with

$$\oplus_\alpha([-\|A\|, \|A\|], d\sigma_\alpha),$$

where $d\sigma_\alpha$ - Stieltjes measures, $\alpha(x) = x$ for all $x \in [-\|A\|, \|A\|]$.

Proof. For the complex H this is the well-known spectral theorem for self-adjoint operators [RS, Theorem VIII.4], [KG, Theorems 9 and 10, Ch. V.2]. Its proof can be modified in standard way to the case of real Hilbert space and symmetric operators. ■

Lemma 3.2.3. *Let H be a real Hilbert space, A a symmetric operator on H and let $H = \oplus_\alpha H_\alpha, A = \oplus_\alpha A_\alpha$ with A_α symmetric oprator on H_α. Suppose that $A_\alpha = \sum_{i=1}^{k} \varepsilon_i^\alpha K_{\alpha,i}^2$ with $\varepsilon_i^\alpha = \pm 1$, $K_{\alpha,i}$ - skew-symmetric operators on H for all $i = 1, ..., k$ k does't depend on α. Then*

$$A = \sum_{i=1}^{2k} \varepsilon_i K_i^2,$$

with $\varepsilon_i = \pm 1$, K_i - skew-symmetric operators on H, $i = 1, ..., 2k$.

Proof. Let $\varepsilon_i = -1$ for $i = 1, ..., k$ and $\varepsilon_i = 1$ for all $i = k+1, ..., 2k$ and put $K_i = \oplus K'_{\alpha,i}$, $i = 1, ..., 2k$, where

$$K'_{\alpha,i} = \begin{cases} K_{\alpha,i} & \text{if } \varepsilon_i^\alpha = -1, \\ 0 & \text{if } \varepsilon_i^\alpha = +1, \end{cases}$$

for $i = 1, ..., k$ and

$$K'_{\alpha,i} = \begin{cases} 0 & \text{if } \varepsilon^{\alpha}_{i-k} = -1, \\ K_{\alpha,i-k} & \text{if } \varepsilon^{\alpha}_{i-k} = +1, \end{cases}$$

for $i = k+1, ..., 2k$.

It clear that K_i are skew-symmetric operators on H and $A = \sum_{i=1}^{2k} \varepsilon_i K_i^2$. ■

Lemma 3.2.4. *Let (X, m) be a measure space, f an essentially bounded real function on X such that*

(i) there exists disjoint measurable subsets S_1, S_2, N of X with

$$X = S_1 \bigcup S_2 \bigcup N;$$

(ii) there is a a measure-preserving transformation α of period 2 on X with

$$\alpha(S_1) = S_2, \alpha(S_2) = S_1, \ \alpha(N) = N;$$

(iii) $f(\alpha(x)) = f(x)$ for all $x \in X$, and $f(x) = 0$ for $x \in N$.

Then the operator M_f on $L_2(X, m)$ can be represented as

$$M_f = -K_1^2 + K_2^2 \ ,$$

where K_1, K_2 are skew-symmetric bounded operators on $L_2(X, m)$.

Proof. First suppose that $f(x) \geq 0$, $x \in X$. Then putting $K_2 = 0$ and for $\varphi \in L_2(X, m)$

$$K_1\varphi(x) = \begin{cases} \sqrt{f(x)}\ \varphi(\alpha(x)) & , \text{ if } x \in S_1, \\ -\sqrt{f(x)}\ \varphi(\alpha(x)) & , \text{ if } x \in S_2, \\ 0 & , \text{ if } x \in N, \end{cases}$$

we obtain that $M_f = -K_1^2 = -K_1^2 + K_2^2$.

In the general case $f = f_+ - f_-$, $f_+ \geq 0, f_- \geq 0$. The positive function f_+ also satisfies the conditions (i) - (iii) with

$$S_1 \bigcap \{f(x) > 0\},\ S_2 \bigcap \{f(x) > 0\},\ N \bigcup \{f(x) \leq 0\},$$

instead of S_1, S_2, N respectively.
Similarly

$$S_1 \bigcap \{f(x) < 0\},\ S_2 \bigcap \{f(x) < 0\},\ N \bigcup \{f(x) \geq 0\}$$

satisfy (i) - (iii) for f_-. ■

Lemma 3.2.5. *Let $d\sigma$ be a (Stieltjes) measure on $[a,b]$, with σ - monotone increasing continuous function. Then any real function $f(x)$ on $[a,b]$, essentially bounded with respect to $d\sigma$, can be decomposed as $f = f_1 + f_2 + f_3$, where each f_i satisfies the conditions of Lemma 3.2.4.*

Proof. Passing if necessary to the set

$$E = [a,b] \setminus \{x \in [a,b] \mid \exists y \in [a,b] : y < x,\ \sigma(y) = \sigma(x),\ \text{or } \sigma(x) = b\},$$

we may assume that σ is strictly monotone. Let $[a,b] = I_1 \bigcup I_2 \bigcup I_3$ be a partition of $[a,b]$ into the sets I_1, I_2, I_3 with

$$d\sigma(I_1) = d\sigma(I_2) = d\sigma(I_3)$$

and $I_1 < I_2 < I_3$ (i.e. $x_1 < x_2 < x_3$ for $x_i \in I_i, i = 1,2,3$). Such a partition exists in viev of the continuity of σ. Define the mappings $\beta_{ij} : I_i \to I_j$ by

$$\beta_{ij}(x) = \sigma^{-1}\left(\sigma(x) + \frac{\sigma(b) - \sigma(a)}{3}(i-j)\right), \quad x \in I_i,$$

and the mappings $\beta_i : [a,b] \to I_i$ by $\beta_i(x) = \beta_{ji}(x)$ for $x \in I_j$. Put for $i = 1,2,3$

$$f_i(x) = \chi_{[a,b]\setminus I_i}(x)\left(\frac{f(\beta_1(x)) + f(\beta_2(x)) + f(\beta_3(x))}{2} - f(\beta_i(x))\right),$$

and

$$\alpha_i(x) = \begin{cases} x & , \text{ for } \quad x \in I_i, \\ \beta_j(x) & , \text{ for } \quad x \in I_k, \\ \beta_k(x) & , \text{ for } \quad x \in I_j, \end{cases}$$

where (i,j,k) is an arbitrary permutation of $(1,2,3)$. Then it is clear that $f = f_1 + f_2 + f_3$ and each α_i ia a transformation of $[a,b]$. A straightforward calculation shows that each f_i satisfies the conditions of Lemma 3.2.4 with $\alpha = \alpha_i$, $N = I_i$, $S_1 = I_j$, $S_2 = I_k$. ■

Now the last Lemma 3.2.4 combined with Lemma 3.2.5 implies

Lemma 3.2.6. *Under the conditions of Lemma 3.2.5, given any essentially bounded real function $a(x)$ on $X = [a,b]$, the symmetric operator M_a on $L_2(X, d\sigma)$ can be represented as*

$$M_a = \sum_{i=1}^{6} \varepsilon_i K_i^2$$

with $\varepsilon_i = \pm 1$, K_i - skew-symmetric operators on $L_2(X, d\sigma)$, $i = 1, ..., 6$.

Proof of Theorem 3.2.1.

According to Lemma 3.2.2 $A = \oplus_\gamma A_\gamma$ and each A_γ can be identified with the operator M_α of multiplication by the function $\alpha(x) = x$ on an appropriate real L_2-space $L_2([a,b], d\sigma)$, with $d\sigma$ - a continuous or discrete measure on $[a,b]$. If $d\sigma$ is continuous then $A_\gamma = \sum_{i=1}^{6} \varepsilon_i K_i^2$ by Lemma 3.2.6. If $d\sigma$ is discrete then A_γ is a finite or infinite diagonal matrix, so according to the condition of theorem we may assume that $dim H_\gamma \geq 3$. If $dim H_\gamma$ is infinite or finite and divisible by 3, then $A_\gamma = \oplus_\beta A'_\beta$, where each A'_β is a diagonal 3×3 matrix. Since

$$diag\{p, q, r\} \;=\; diag\{\frac{p+q-r}{2}, \frac{p+q-r}{2}, 0\} +$$

$$+ \, diag\{\frac{p-q+r}{2}, 0, v\frac{p-q+r}{2}\} + diag\{0, \frac{-p+q+r}{2}, \frac{-p+q+r}{2}\}$$

and

$$\begin{pmatrix} u & 0 \\ 0 & u \end{pmatrix} = sgnu \begin{pmatrix} 0 & \sqrt{|u|} \\ -\sqrt{|u|} & 0 \end{pmatrix}^2$$

each A'_β is represented as $\sum_{i=1}^{3} \varepsilon_i K_i^2$, where K_i are skew-symmetric matrices. If $dim H_\gamma$ is finite but not a multiple of 3, then $A_\gamma = B_\gamma + C_\gamma$, where each B_γ and C_γ can be considered as a matrix of order $3k$, $k = 1, 2, 3, ...$). So in the general case $A_\gamma = \sum_{i=1}^{6} \varepsilon_i K_i^2$ if $d\sigma$ is discrete. Now Lemma 3.2.3 implies that $A = \sum_{i=1}^{12} \varepsilon_i K_i^2$. ∎

Now we shall extend the above theorem to Hilbert spaces over the skew field $\mathbb{Q}$ of quaternions.

Theorem 3.2.7. *Let H be a quaternionian Hilbert space, $B(H_\mathbb{Q})$ - the algebra of all bounded $\mathbb{Q}$ - linear operators on $H_\mathbb{Q}$. Then for any self-adjoint operator $A \in B(H_\mathbb{Q})$ there exist 21 skew-adjoint operators $K_1, K_2, ..., K_{21} \in B(H_\mathbb{Q})$ such that*

$$A = \sum_{i=1}^{21} \varepsilon_i K_i^2,$$

where $\varepsilon_i = \pm 1, i = 1, 2, ..., 21$.

Proof. First let $dim H \geq 3$ (dim means the dimension over $\mathbb{Q}$). Then

$$H_\mathbb{Q} = H + iH + jH + kH,$$

where i, j, k are unit quaternions, H is real Hilbert space with real dimension ≥ 3. Therefore any self-adjoint operator $A \in B(H_{\mathbb{Q}})$ can be decomposed as

$$A = A_0 + iA_1 + jA_2 + kA_3,$$

where $A_0, A_1, A_2, A_3 \in B(H)$, A_0 is symmetric and A_1, A_2, A_3 are skew-symmetric. By Theorem 3.2.1 $A_0 = \sum_{i=1}^{12} \varepsilon_i K_i^2$, with K_i skew-symmetric operators from $B(H)$. Further

$$iA_1 = i\mathbf{I}A_1 = \frac{1}{2}[(i\mathbf{I} + A_1)^2 - A_1^2 - (i\mathbf{I})^2],$$

with $(i\mathbf{I} + A_1), A_1,\ i\mathbf{I}$ - skew-adjoint elemts in $B(H_{\mathbb{Q}})$. Therefore $iA_1 + jA_2 + kA_3$ is a sum of 9 squares of skew-adjoint operators with coefficients ± 1, and hence

$$A = \sum_{i=1}^{21} \varepsilon_i K_i^2.$$

Now let $dim H_{\mathbb{Q}} = 1$, i.e. $B(H_{\mathbb{Q}}) = \mathbb{Q}$. Then self-adjoint elements are real numbers and for $a \in \mathbf{R}$:

$$a = -sgna(i\sqrt{|a|})^2.$$

Finally, let $dim H_{\mathbb{Q}} = 2$, i.e. $B(H_{\mathbb{Q}})$ coincides with the algebra $M_2(\mathbb{Q})$ of quaternionian 2×2 matrices. Every self-adjoint matrix from $M_2(\mathbb{Q})$ is a real linear combination of the following 6 matrices

$$\begin{pmatrix} 1 & 0 \\ 0 & 0 \end{pmatrix}, \begin{pmatrix} 0 & 0 \\ 0 & 1 \end{pmatrix}, \begin{pmatrix} 0 & 1 \\ 1 & 0 \end{pmatrix},$$

$$\begin{pmatrix} 0 & i \\ -i & 0 \end{pmatrix}, \begin{pmatrix} 0 & j \\ -j & 0 \end{pmatrix}, \begin{pmatrix} 0 & k \\ -k & 0 \end{pmatrix}.$$

Each of these matrices is a sum of 3 squares of skew-adjoint matrices with coefficients ± 1. For example

$$\begin{pmatrix} 1 & 0 \\ 0 & 0 \end{pmatrix} = -\begin{pmatrix} i & 0 \\ 0 & 0 \end{pmatrix}^2,$$

$$\begin{pmatrix} 0 & 1 \\ 1 & 0 \end{pmatrix} = -\begin{pmatrix} i & 0 \\ 0 & 0 \end{pmatrix}^2 + \begin{pmatrix} 0 & i \\ i & 0 \end{pmatrix}^2 - \begin{pmatrix} i & i \\ i & 0 \end{pmatrix}^2,$$

$$\begin{pmatrix} 0 & i \\ -i & 0 \end{pmatrix} = \begin{pmatrix} j & k \\ k & 0 \end{pmatrix}^2 - \begin{pmatrix} j & 0 \\ 0 & 0 \end{pmatrix}^2 - \begin{pmatrix} 0 & k \\ k & 0 \end{pmatrix}^2, \; etc.$$

Therefore any self-adjoint matrix from $M_2(\mathbb{Q})$ may be represented as a sum of 18 squares of skew-adjoint matrices. ■

Remark 1. It is clear that unlike Theorem 3.2.7 the assertion of Theorem 3.2.1 is not true for real Hilbert spaces of dimension 1 and 2.

Remark 2. The analogue of Theorem 3.2.7 is trivial for complex Hilbert spaces. Indeed, for any self-adjoint operator A on a complex Hilbert space, we have $A = A_+ - A_-$, where $A_+, A_- \geq 0$. Therefore

$$A = -(i\sqrt{A_+})^2 + (i\sqrt{A_-})^2.$$

The above results show that the mentioned theorem of Baxter [Bax] is true also for prime (but not simple in general) algebras $B(H)$, where H is a real, complex or quaternionian Hilbert space. But for general prime algebras it is not true as the following example shows.

Example 3.2.8. Let $R = R(x_1, x_2, ...)$ be the free non commutative algebra over $\mathbf{R}$ with generators $x_1, x_2, ...$ Define the involution on R putting $x_i^* = x_i$ for generators x_i, $i = 1, 2, ...,$ then extending it to the whole R in the standart way. We obtain a prime algebra with involution of the first kind. By definition skew-symmetric elements of R must contain at least two generators. Therefore, symmetric elements x_i cannot be represented as sums of squares of skew-symmetric elements from R.

We now turn to skew-symmetric operators on real a Hilbert space. If R is a simple ring with involution I.N.Herstein [He 1] asked whether every skew element in R is a sum of commutators of skew elements. A negative answer to this question was given by P.H.Lee [L]. Nevertheless we shall prove that Herstein's conjecture is true for the algebra $B(H)$.

Theorem 3.2.9. *Let H be a real Hilbert space, $dim H \geq 3$. Given any skew-symmitric operator $A \in B(H)$ there are two skew-symmetric operators $B, C \in B(H)$ such that $A = [B, C]$.*

Proof. First let $dim H = n < \infty$. It is well know that in an appropriate basis of H the skew-symmetric operator A has the matrix

$$A = \oplus_{k=1}^{p} \begin{pmatrix} 0 & a_k \\ -a_k & 0 \end{pmatrix} \oplus O_q$$

where O_q is the zero matric of order q, and $2p + q = n$. Consider the following two matrices

$$B = O_1 \oplus \oplus_{k=1}^{p-1} \begin{pmatrix} 0 & \sqrt{k} \\ -\sqrt{k} & 0 \end{pmatrix} \oplus O_{q+1},$$

$$C = C_{2p} \oplus O_q,$$

where C_{2p} is the skew-symmetric matrix whose entries in the diagonal second above the main diagonal are γ_i, i.e. $c_{i,i+2} = \gamma_i$, $i = 1, 2, ..., 2p-2$; entries in the diagonal second below the main diagonal are $-\gamma_i$, i.e. $c_{i+2,i} = -\gamma_i$, $i = 1, 2, ..., 2p - 2$; all other entries being zero. Now if we put $\gamma_{2m} = \sqrt{m} \cdot a_{m+1}$, $\gamma_{2m+1} = \sqrt{m-1} \cdot a_{m-1}$, $m = 1, ..., p-1$, the straitforward computations show that $A = BC - CB = [B, C]$.

Now suppose that $dim H = \infty$ and without loss of generality we may assume that H is separable. Since A is skew-symmetric there exists an orthonormal basis of H such that A is given by an infinite matrix

$$A = \begin{pmatrix} 0 & a_1 & 0 & 0 & \cdots \\ -a_1 & 0 & a_2 & 0 & \cdots \\ 0 & -a_2 & 0 & a_3 & \cdots \\ 0 & 0 & -a_3 & 0 & \cdots \\ 0 & 0 & 0 & -a_4 & \cdots \\ . & . & . & . & \cdots \end{pmatrix}$$

(see [Ha 2, Problem 36]). Now following V.Yu.Sadovskij [Sad] put

$$B = \oplus_{k=1}^{\infty} \begin{pmatrix} 0 & 0 & 0 & 0 \\ 0 & 0 & 0 & 1 \\ 0 & 0 & 0 & 0 \\ 0 & -1 & 0 & 0 \end{pmatrix}, \qquad C = (c_{k,l})_{k,l=1}^{\infty},$$

where

$$\begin{aligned} c_{4i-3,4i} &= a_{4i-3}, & c_{4i,4i-3} &= -a_{4i-3}, \\ c_{4i-2,4i-1} &= a_{4i-1}, & c_{4i-1,4i-2} &= -a_{4i-1}, \\ c_{4i-2,4i+1} &= -a_{4i}, & c_{4i+1,4i-2} &= a_{4i}, \\ c_{4i-1,4i} &= -a_{4i-2}, & c_{4i,4i-1} &= a_{4i-2}, \end{aligned}$$

all other entries being zero. It is clear that B and C are skew-symmetric and direct calculations show that $A = [B, C]$. Finally note that each of the matrices A, B, C has at most two non zero entries in each row and each column. Such a matrix corresponds to a bounded linear operator on H if and only if all its entries are uniformly bounded [Sad]. Therefore B is bounded and the boundedness of A implies that C is also bounded. ∎

Corollary 3.2.10. *Each symmetric operator* $x \in B(H)$ *is of the form*

$$x = \sum_{i=1}^{12} \pm [x_i, y_i]^2,$$

where x_i, y_i *skew-symmetric operators from* $B(H)$, $i = 1, 2, ..., 12$.

Corollary 3.2.11. *Let* $H_{\mathbb{Q}}$ *be a quaternionian Hilbert space. Then any skew-adjoint* $A \in B(H_{\mathbb{Q}})$ *is a finite sum of commutators* $[B, C]$ *with* B, C *- skew-adjoint operators from* $B(H_{\mathbb{Q}})$.

Proof. Similar to the proof of Theorem 3.2.7 we have

$$A = A_0 + iA_1 + jA_2 + kA_3,$$

where $A_0, A_1, A_2, A_3 \in B(H)$, A_0 - skew-symmetric, A_1, A_2, A_3 are symmetric operators. If $dim H_{\mathbb{Q}} > 2$, then $dim H > 2$ and Theorem 3.2.9 implies that $A = [B_0, C_0]$ with B_0, C_0 skew-symmetric operators from $B(H)$. Therefore

$$A = \left[B_0, C_0\right] + \left[j\mathbf{1}, \frac{k}{2}A_1\right] + \left[k\mathbf{1}, \frac{i}{2}A_2\right] + \left[i\mathbf{1}, \frac{j}{2}A_3\right],$$

and the assertion is proved for $dim H_{\mathbb{Q}} > 2$.

If $dim H_{\mathbb{Q}} = 1$, then $B(H_{\mathbb{Q}}) = \mathbb{Q}$, and every skew element is of the form $q = ib + jc + kd$, where b, c, d are real numbers. Therefore

$$2q = [j, kb] + [k, ic] + [k, id].$$

If $dim H_{\mathbb{Q}} = 2$, then $B(H_{\mathbb{Q}}) = M_2(\mathbb{Q})$, and each skew-adjoint matrix is a real linear combination of the following matrices

$$\begin{pmatrix} i & 0 \\ 0 & 0 \end{pmatrix}, \begin{pmatrix} j & 0 \\ 0 & 0 \end{pmatrix}, \begin{pmatrix} k & 0 \\ 0 & 0 \end{pmatrix}, \begin{pmatrix} 0 & 0 \\ 0 & i \end{pmatrix},$$

$$\begin{pmatrix} 0 & 0 \\ 0 & j \end{pmatrix}, \begin{pmatrix} 0 & 0 \\ 0 & k \end{pmatrix}, \begin{pmatrix} 0 & 1 \\ -1 & 0 \end{pmatrix}, \begin{pmatrix} 0 & i \\ i & 0 \end{pmatrix},$$

$$\begin{pmatrix} 0 & j \\ j & 0 \end{pmatrix}, \begin{pmatrix} 0 & k \\ k & 0 \end{pmatrix},$$

each of the above matrices is the commutator of two skew-adjoint matrices, e.g.

$$\begin{pmatrix} i & 0 \\ 0 & 0 \end{pmatrix} = \frac{1}{2}\left[\begin{pmatrix} j & 0 \\ 0 & 0 \end{pmatrix}, \begin{pmatrix} k & 0 \\ 0 & 0 \end{pmatrix}\right],$$

$$\begin{pmatrix} 0 & 1 \\ -1 & 0 \end{pmatrix} = \left[\begin{pmatrix} 0 & i \\ i & 0 \end{pmatrix}, \begin{pmatrix} i & 0 \\ 0 & 0 \end{pmatrix}\right],$$

$$\begin{pmatrix} 0 & i \\ i & 0 \end{pmatrix} = \left[\begin{pmatrix} j & 0 \\ 0 & 0 \end{pmatrix}, \begin{pmatrix} 0 & k \\ k & 0 \end{pmatrix}\right], \text{ etc. } \blacksquare$$

Remark 1. The last corollary can be stengthened by showing that in fact any skew-adjoint operator on a quaternionian Hilbert space is a single commutators of skew-adjoint operators [Sad].

Remark 2. It is clear that for $dim H = 2$, i.e. for real 2×2 matrices $M_2(\mathbf{R})$, skew matrices cannot be represented as sums of commutators of skew matrices, because any two skew matrices B, C commute, i.e. $[B, C] = 0$.

Remark 3. For complex Hilbert spaces the picture is completely different. In the finite dimensional case the skew-adjoint operator $i\mathbf{I}$ cannot be represented as a sum of commutators (trace argument). In the infinite dimensional complex Hilbert space H a result of Halmos [Ha 1] (see also [Ha 2, Problem 186]), implies that every skew-adjoint oprator from $B(H)$ is a sum of 4 commutators of skew-adjoint operators from $B(H)$.

Proposition 3.2.12. *Let $\mathfrak{R}$ be a real factor. Then*

$$\mathfrak{R}_k = [\mathfrak{R}_k, \mathfrak{R}_k] + [\mathfrak{R}_s, \mathfrak{R}_s].$$

Proof. Let $M = \mathfrak{R} + i\mathfrak{R}$ be the enveloping (complex) W^*-algebra. Then M is a factor and its type coincides with the type of $\mathfrak{R}$ (see Theorem 1.3.2). Consider two cases separetely:

(i) $\mathfrak{R}$ is finite,

(ii) $\mathfrak{R}$ is properly infinite.

(i) Let $\mathfrak{R}$ be a finite real factor. By Corollary 1.2.4 it possesses a finite normal trace τ. By Theorem 1.2.9 τ is extended by linearity to whole M and it is a finite normal trace, which we denote by τ_0. Positivity of τ_0 implies that given any $\tau_0(-x) = \tau_0(x^*) = \overline{\tau_0(x)}$, for $x \in M_k$. In particular if $x \in \mathfrak{R}_k$, then $\tau(-x) = \tau(x)$, i.e. $\tau_0(x) = \tau(x) = 0$ for all $x \in \mathfrak{R}$. Now [FH, Theorem 3.2] shows that x is a sum of 10 commutators, i.e.

$$x = \sum_{j=1}^{10} [a_j, b_j], \quad a_j, b_j \in M.$$

Since $M = \Re + i\Re$, we can expess each a_j, b_j as

$$a_j = q_j + ir_j,\ b_j = s_j + it_j; \quad q_j, r_j, s_j, t_j \in \Re,\ j = 1, ..., 10.$$

Therefore

$$x = \sum_{j=1}^{10} [q_j + ir_j, s_j + it_j] =$$

$$\sum_{j=1}^{10} ([q_j, s_j] - [r_j, t_j]) + i \cdot \sum_{j=1}^{10} ([r_j, s_j] + [q_j, t_j]).$$

Since $x \in \Re$ and $\Re \bigcap i\Re = \{0\}$, it follows that

$$x = \sum_{j=1}^{10} ([q_j, s_j] - [r_j, t_j]),$$

i.e. x is a sum of 20 commutators of elements from $\Re$, i.e. $\Re_k \subset [\Re, \Re]$.

(ii) Now let $\Re$ be a properly infinite real factor. Then Theorem 1.3.2 implies that M is also properly infinite and [Su, Theorem 2] gives that $M = [M, M]$, i.e.

$$\Re + i\Re = [\Re, \Re] + i[\Re, \Re].$$

Since $\Re \bigcap i\Re = \{0\}$ it follows that $\Re = [\Re, \Re]$, in particular $\Re_k \subset [\Re, \Re]$.

Thus in both cases $\Re_k \subset [\Re, \Re]$. Since $\Re$ is direct sum $\Re = \Re_s \oplus \Re_k$ it follows that

$$\Re_k \subset [\Re_s, \Re_s] + [\Re_k, \Re_k] + [\Re_s, \Re_k] + [\Re_k, \Re_s].$$

But $[\Re_k, \Re_s] \subset \Re_s$, $[\Re_s, \Re_k] \subset \Re_s$, and $\Re_s \bigcap \Re_k = \{0\}$. Therefore

$$\Re_k = [\Re_s, \Re_s] + [\Re_k, \Re_k]. \qquad \blacksquare$$

3.3. Commutators of skew elements in real factors.

Theorem 3.3.1. *Let $\Re$ be a real factor not of type I_1 and I_2. Then any symmetric element $a \in \Re_s$ is a finite sum of squares of skew commutators with coefficients ± 1, i.e.*

$$a = \sum_{i=1}^{N} \pm [x_i, y_i]^2,$$

where $x_i, y_i \in \Re_k$, $i = 1, ..., N$; moreover $N \leq 25$.

Proof. If $\Re$ has type I, then $\Re = B(H)$, where H is a real or quaternionian Hilbert space and Theorems 3.2.1 and 3.2.7 imply that

$$a = \sum_{i=1}^{21} \pm [x_i, y_i]^2 \quad \text{with} \quad x_i, y_i \in \Re_k, \ i = 1, ..., 21.$$

Therefore we have to consider only continuous factors (i.e. of type II and III). In such factors there exist three mutually orthogonal and mutually equivalent projections with sum **I**. As in the case of complex factors [T 1, Ch. IV, Proposition 1.8] the real factor $\Re$ is *-isomorphic to the tensor product $\Re^0 \otimes M_3(\mathbf{R})$, where $\Re^0$ is a real factor of the same type as $\Re$, and $M_3(\mathbf{R})$ is the algebra of all real 3×3 matrices. Elements of $\Re^0 \otimes M_3(\mathbf{R})$ can be identified with 3×3 matrices with elements from $\Re^0$. So from now on we consider elements of $\Re$ as 3×3 matrices over a real factor $\Re^0$, i.e. $\Re = M_3(\Re^0)$. Let $a = a^* \in \Re_s$, then

$$a = \begin{pmatrix} d_1 & m_1^* & m_2^* \\ m_1 & d_2 & m_3^* \\ m_2 & m_3 & d_3 \end{pmatrix}$$

where $d_i \in \Re^0_s$, $m_i \in \Re^0$, $i = 1, 2, 3$.

If $m_i = b_i u_i$ is the polar decomposition of the operator $m_i \in \Re^0$, then by [S 5, Lemma 1.1] the elements b_i, u_i are also in the real factor $\Re^0$. In the weakly-closed real abelian subalgebra generated by the symmetric element (i.e. self-adjoint operator) b_i there exists a symmetric element $c_i \in \Re_s^0$ such that

$$(2\|b_i\|)^{-1}b_i = c_i\sqrt{1-c_i^2} \qquad \text{(cf. [PT, Lemma 2.2]).}$$

Put $w_i = c_i u_i$, $i = 1, 2, 3$. Then as in [PT, Lemma 2.2] the elements

$$e_i = \begin{pmatrix} w_1^* w_1 & w_1^*\sqrt{1-w_1w_1^*} & 0 \\ \sqrt{1-w_1w_1^*}w_1 & 1-w_1w_1^* & 0 \\ 0 & 0 & 0 \end{pmatrix} =$$

$$= \frac{1}{2\|b_1\|}\begin{pmatrix} 2\|b_1\|w_1^*w_1 & m_1^* & 0 \\ m_1 & 2\|b_1\|(1-w_1w_1^*) & 0 \\ 0 & 0 & 0 \end{pmatrix}$$

$$e_2 = \begin{pmatrix} w_2^* w_2 & 0 & w_2^*\sqrt{1-w_2w_2^*} \\ 0 & 0 & 0 \\ \sqrt{1-w_2w_2^*}w_2 & 0 & 1-w_2w_2^* \end{pmatrix} =$$

$$= \frac{1}{2\|b_2\|}\begin{pmatrix} 2\|b_2\|w_2^*w_2 & 0 & m_2^* \\ 0 & 0 & 0 \\ m_2 & 2\|b_2\|(1-w_2w_2^*) & \end{pmatrix}$$

$$e_3 = \begin{pmatrix} 0 & 0 & 0 \\ 0 & w_3^* w_3 & w_3^*\sqrt{1-w_3w_3^*} \\ 0 & \sqrt{1-w_3w_3^*}w_3 & 1-w_3w_3^* \end{pmatrix} =$$

$$= \frac{1}{2\|b_3\|}\begin{pmatrix} 0 & 0 & 0 \\ 0 & 2\|b_3\|w_3^* w_3 & m_3^* \\ 0 & m_3 & 2\|b_3\|(1 - w_3 w_3^*) \end{pmatrix},$$

are projections in $\Re$, i.e. $e_i^* = e_i = e_i^2$, $i = 1,2,3,$ such that the element

$$d = a - 2\sum_{i=1}^{3} \|b_i\| e_i$$

is a diagonal matrix. The operator $d + \|d\|\mathbf{I}$ is positive and therefore has the form

$$diag\{\ell_1, \ell_2, \ell_3\} = \begin{pmatrix} \ell_1 & 0 & 0 \\ 0 & \ell_2 & 0 \\ 0 & 0 & \ell_3 \end{pmatrix}$$

with $\ell_1, \ell_2, \ell_3 \in (\Re_s^0)^+$, $i = 1,2,3$. We can decompose this operator into the sum of 3 matrices, each having only one non zero positive element on its diagonal. Consider one of such matrices, e.g. $diag\{\ell, 0, 0\}$, where $\ell \in \Re_s^0, \ell \geq 0$. We have

$$diag\{\ell, 0, 0\} = \frac{1}{2}\Big(diag\{\ell, -\ell, 0\} + diag\{\ell, 0, -\ell\} + diag\{0, \ell, \ell\}\Big).$$

Let us show that $diag\{\ell, -\ell, 0\}$ is a linear combination of three projections. Without loss of generality we may assume that $\|\ell\| \leq 1$. Consider the operator

$$f = (\mathbf{I} + 2\ell)^{-1}(\ell^2 + 2\ell) \in \Re_s^0, \quad f \geq 0,$$

and put

$$g = g(f) = \frac{1}{2}(\mathbf{I} + \sqrt{f^2 - f + \mathbf{I}}) \in \Re_s^0.$$

It is easy to see that

$$\sqrt{f(\mathbf{I} - f)} = 2\sqrt{g(\mathbf{I} - g)} \quad \text{and} \quad f - 2g = f - \mathbf{I} - \sqrt{f^2 - f + \mathbf{I}}.$$

Therefore

$$diag\{\ell,-\ell,0\} = \begin{pmatrix} f & \sqrt{f(\mathbf{1}-f)} & 0 \\ \sqrt{f(\mathbf{1}-f)} & \mathbf{1}-f & 0 \\ 0 & 0 & 0 \end{pmatrix} -$$

$$2\cdot\begin{pmatrix} g & \sqrt{g(\mathbf{1}-g)} & 0 \\ \sqrt{g(\mathbf{1}-g)} & \mathbf{1}-g & 0 \\ 0 & 0 & 0 \end{pmatrix} + diag\{0,1,0\}.$$

and a straightforward calculation shows that each of three matrices on the right hand is a projection. Similarly, the matrix

$$diag\{\ell,0,-\ell\}$$

is also a linear combination of three projections. Finally, the matrix

$$diag\{0,\ell,\ell\} = \begin{pmatrix} 0 & 0 & 0 \\ 0 & 0 & \sqrt{\ell} \\ 0 & -\sqrt{\ell} & 0 \end{pmatrix}^2 =$$

$$= \left[\begin{pmatrix} 0 & -1 & 0 \\ 1 & 0 & 0 \\ 0 & 0 & 0 \end{pmatrix}, \begin{pmatrix} 0 & 0 & \sqrt{\ell} \\ 0 & 0 & 0 \\ -\sqrt{\ell} & 0 & 0 \end{pmatrix}\right]^2$$

is the square of the commutator of two skew operators. Thus each of 3 matrices with one non zero positive element on the diagonal is a linear combination of 6 projections and one square of a skew commutator. Taking into account the projections e_1, e_2, e_3 and $\|d\|\mathbf{1}$, above, we obtain that every $a = a^* \in \mathfrak{R} = \mathfrak{R}^0 \otimes M_3(\mathbf{R})$ is a linear combination of 22 projections and 3 skew commutators. Below in Lemma 3.3.2 we shall prove that each projection in a continuous factor is the negative of the square of a skew commutator. Therefore

$$a = \sum_{i=1}^{25} \pm[x_i, y_i]^2, \qquad x_i, y_i \in \mathfrak{R}_k, \quad i = 1, \ldots, 25. \quad \blacksquare$$

Lemma 3.3.2. *Let* $\mathfrak{R}$ *be a continuous real factor,* $\mathcal{P}_{\mathfrak{R}}$ *its projection lattice. Then any projection* $p \in \mathfrak{R}$ *is of the form* $p = -[x,y]^2$*, with* $x, y \in \mathfrak{R}_k$*. In particular,* $\mathcal{P}_{\mathfrak{R}} \subset \overline{[\mathfrak{R}_k, \mathfrak{R}_k]}$.

Proof. Considering the algebra $p\mathfrak{R}p$, we may assume without loss of generality that $p = \mathbf{1}$. By [ASS, Theorem 6.9] applied to the JW-algebra $\mathfrak{R}_s$, there exist four mutually orthogonal projections $e_1, e_2, e_3, e_4 \in \mathcal{P}_{\mathfrak{R}}$, $\mathbf{1} = e_1 + e_2 + e_3 + e_4$, any two of which are exchanged via symmetry in $\mathfrak{R}$, i.e.

$$e_i = s_i e_1 s_i, \quad s_i \in \mathfrak{R}_s, \quad s_i^2 = \mathbf{1}, \quad i = 2,3,4.$$

Put

$$x = s_2 e_1 s_3 - s_3 e_1 s_2, \quad y = e_1 s_3 + s_2 e_1 s_4 - s_3 e_1 - s_4 e_1 s_2.$$

It is clear that $x, y \in \mathfrak{R}_k$ and in view of orthogonality of projections e_i and the above relations we have

$$[x,y] = e_1 s_2 + s_4 e_1 s_3 - s_2 e_1 - s_3 e_1 s_4,$$

and

$$[x,y]^2 = -e_1 - e_2 - e_3 - e_4 = -\mathbf{1}. \quad \blacksquare$$

Now Theorem 3.3.1 implies

Corollary 3.3.3. *In a real factor* $\mathfrak{R}$ *not of type* I_1 *and* I_2 *one has*

$$\mathfrak{R}_s \subset \overline{[\mathfrak{R}_k, \mathfrak{R}_k]}.$$

Remark 1. Theorem 3.3.1 extends to prime operator algebras a result of Chuang and Lee [CL]: in simple algebras with involution of the first kind $\mathfrak{R}_s \subset [\mathfrak{R}_k, \mathfrak{R}_k] \cdot [\mathfrak{R}_k, \mathfrak{R}_k]$, which gives a positive solution of Martindale's conjecture [CM].

In [He 1] Herstein asked whether $\mathfrak{R}_k = [\mathfrak{R}_k, \mathfrak{R}_k]$ in a simple ring $\mathfrak{R}$ with involution of first kind if $char\mathfrak{R} \neq 2$, $dim\mathfrak{R} > 4$ (or $Z = \{0\}$). A negative answer to this question was given by Lee [L], who presented an example of division rings with involution in which some skew elements cannot be expressed as a sum of commutators, i.e. $\mathfrak{R}_k \not\subset [\mathfrak{R}_k, \mathfrak{R}_k]$. Nevertheless in the case of operator algebras it was proved in Section 3.2, that if $\mathfrak{R}$ is a real factor of type $\mathrm{I}_n (3 \leq n \leq \infty)$, then $\mathfrak{R}_k = [\mathfrak{R}_k, \mathfrak{R}_k]$. We are now in a position to prove that Herstein conjecture is true even for prime (or topologically simple) operator algebras.

Theorem 3.3.4. *Let $\mathfrak{R}$ be a real factor different from $M_2(\mathbf{R})$. Then*

$$\mathfrak{R}_k = [\mathfrak{R}_k, \mathfrak{R}_k].$$

Proof. According to Theorem 3.2.9 and Corollary 3.2.11 the assertion is true for real factors of type I except $M_2(\mathbf{R})$. Therefore it is sufficient to consider continuous real factors, which can be divided into the following two subcases:

(i) $\mathfrak{R}$ is of type II_1 ;

(ii) $\mathfrak{R}$ is properly infinite, i.e. of type II_∞ or III.

In the case (i) $\mathfrak{R}$ is an infinite dimensional simple algebra with involution of first kind (Proposition 3.1.3) and [Bax, Lemma 10] implies that $[\mathfrak{R}_k, \mathfrak{R}_k] = [\mathfrak{R}_s, \mathfrak{R}_s]$. But from Proposition 3.2.12 we have that

$$\mathfrak{R}_k = [\mathfrak{R}_k, \mathfrak{R}_k] + [\mathfrak{R}_s, \mathfrak{R}_s],$$

and therefore

$$\mathfrak{R}_k = [\mathfrak{R}_k, \mathfrak{R}_k] + [\mathfrak{R}_s, \mathfrak{R}_s] = [\mathfrak{R}_k, \mathfrak{R}_k] + [\mathfrak{R}_k, \mathfrak{R}_k] = [\mathfrak{R}_k, \mathfrak{R}_k],$$

which proves the statement in the case (i). The proof of the theorem in the case (ii) follows from the next more explicit result.

Lemma 3.3.5. *Let $\mathfrak{R}^0$ be a properly infinite real factor. Then each skew element in $\mathfrak{R}^0$ is a sum of three skew commutators.*

Proof. Since $\mathfrak{R}^0$ is properly infinite there exist a projection $e \in \mathfrak{R}^0$ such that $e \sim \mathbf{I} - e$ (cf. [T 1, Ch. V., Proposition 1.36]) and also a family of mutually orthogonal projections $e_1, e_2, ..., e_n, ...$such that $e \sim e_i$ for all $i = 1, 2, ...,$ and $e_1 + e_2 + \cdots = \mathbf{I} - e$ (cf. [PT, §3]). As in Theorem 3.3.1 this implies that

$$\mathfrak{R}^0 \cong M_2(\mathfrak{R}) = \mathfrak{R} \otimes M_2(\mathbf{R}),$$

where $\mathfrak{R} \cong e\mathfrak{R}^0 e$. Thus each element $a \in \mathfrak{R}^0$ is a 2×2 matrix

$$a = \begin{pmatrix} x & y \\ z & u \end{pmatrix}$$

with $x, y, z, u \in \mathfrak{R}$. If we put

$$b = b(a) = \begin{pmatrix} y & -x \\ u & -z \end{pmatrix}, \quad c = \begin{pmatrix} 0 & 1 \\ -1 & 0 \end{pmatrix}$$

then it is clear that $a = bc$, $c \in \mathfrak{R}^0_k$, $b = b(a) \in \mathfrak{R}^0$, i.e. each element of $\mathfrak{R}^0$ can be decomposed into the product of the skew element c and the element $b(a)$.

Now consider an arbitrary skew element in $\mathfrak{R}^0$:

$$\begin{pmatrix} k & m^* \\ -m & \ell \end{pmatrix}$$

with $k, \ell \in \mathfrak{R}_k, m \in \mathfrak{R}$. If $-m = bc$ is the decomposition mentioned above with $b = b(-m) \in \mathfrak{R}$, $c \in \mathfrak{R}_k$, then it is clear that

$$\begin{pmatrix} 0 & m^* \\ -m & 0 \end{pmatrix} = \begin{pmatrix} 0 & cb^* \\ bc & 0 \end{pmatrix} = \left[\begin{pmatrix} c & 0 \\ 0 & 0 \end{pmatrix}, \begin{pmatrix} 0 & b^* \\ -b & 0 \end{pmatrix} \right]$$

is a commutator of two skew elements from $\mathfrak{R}^0$. Therefore it is sufficient to represent as skew commutators the matrices

$$\begin{pmatrix} k & 0 \\ 0 & 0 \end{pmatrix} \quad \text{and} \quad \begin{pmatrix} 0 & 0 \\ 0 & \ell \end{pmatrix}$$

with $k, \ell \in \mathfrak{R}_k$. Consider for example the first one.

Since

$$\mathbf{1} - e = e_1 + e_2 + \cdots, \quad e_j \sim e_i, \quad e_j \perp e_i, \quad i \neq j, \ i,j = 1,2,...,$$

similar to [PT, Ch. 3] elements from $\mathfrak{R}^0$ can be represented as infinite matrices with elements from $e\mathfrak{R}^0 e = \mathfrak{R}$, i.e. $\mathfrak{R}^0 \cong M_\infty(\mathfrak{R})$.

In this representation the element

$$\begin{pmatrix} k & 0 \\ 0 & 0 \end{pmatrix} \quad \in \mathfrak{R}^0 \ = \ M_2(\mathfrak{R})$$

corresponds to the infinite matrix

$$\begin{pmatrix} k & 0 & 0 & \dots \\ 0 & 0 & 0 & \dots \\ 0 & 0 & 0 & \dots \\ . & . & . & \dots \end{pmatrix} =$$

$$= \frac{1}{2}\left[\begin{pmatrix} 0 & 1 & 0 & 0 & \dots \\ -1 & 0 & 1 & 0 & \dots \\ 0 & -1 & 0 & 1 & \dots \\ 0 & 0 & -1 & 0 & \dots \\ . & . & . & . & \dots \end{pmatrix}, \begin{pmatrix} 0 & k & 0 & 0 & \dots \\ k & 0 & k & 0 & \dots \\ 0 & k & 0 & k & \dots \\ 0 & 0 & k & 0 & \dots \\ . & . & . & . & \dots \end{pmatrix}\right],$$

which is a skew commutator. The proof of Lemma 3.3.5 and therefore of Theorem 3.3.4 is complete. ∎

If in the above proof of Lemma 3.3.5 we consider the polar decomposition $m = u|m|$, instead of the decomposition of $-m$ as $-m = bc$ with $b = b(-m) \in \mathfrak{R}$, $c \in \mathfrak{R}_k$, we obtain

$$\begin{pmatrix} 0 & m^* \\ -m & 0 \end{pmatrix} = \begin{pmatrix} 0 & |m|u^* \\ -u|m| & 0 \end{pmatrix} = \left[\begin{pmatrix} |m| & 0 \\ 0 & 0 \end{pmatrix}, \begin{pmatrix} 0 & u^* \\ u & 0 \end{pmatrix}\right],$$

i.e. a representation as a commutator of two symmetric elements.

Further the matrix

$$\begin{pmatrix} k & 0 & 0 & \cdots \\ 0 & 0 & 0 & \cdots \\ 0 & 0 & 0 & \cdots \\ . & . & . & \cdots \end{pmatrix} =$$

$$= \frac{1}{2}\left[\begin{pmatrix} 0 & 1 & 0 & 0 & \cdots \\ 1 & 0 & 1 & 0 & \cdots \\ 0 & 1 & 0 & 1 & \cdots \\ 0 & 0 & 1 & 0 & \cdots \\ . & . & . & . & \cdots \end{pmatrix}, \begin{pmatrix} 0 & k & 0 & 0 & \cdots \\ -k & 0 & k & 0 & \cdots \\ 0 & -k & 0 & k & \cdots \\ 0 & 0 & -k & 0 & \cdots \\ . & . & . & . & \cdots \end{pmatrix}\right]$$

is also a commutator of two symmetric matrices. Therefore similar to the proof of Lemma 3.3.5 we obtain the following.

Lemma 3.3.6. *Let* $\mathfrak{R}^0$ *be a properly infinite real factor. Then each skew element in* $\mathfrak{R}^0$ *ia a sum of three commutators of symmetric elements.*

Now the last lemma enables us to prove the following result.

Theorem 3.3.7. *Let* $\mathfrak{R}$ *be a real factor different from the quaternions* $\mathbb{Q}$. *Then* $\mathfrak{R}_k = [\mathfrak{R}_s, \mathfrak{R}_s]$, *i.e. each skew element is a finite sum of commutators of symmetric elements.*

Proof. If $\mathfrak{R}$ is properly infinite, then Lemma 3.3.6. gives the proof. If $\mathfrak{R}$ is of type II_1 or I_n $(n < \infty)$ then it is a simple algebra (see Theorem 3.1.5) and therefore if its dimension is < 16 over $\mathbf{R}$ then by [Bax, Lemma 10] and Theorem 3.3.4

$$[\mathfrak{R}_s, \mathfrak{R}_s] = [\mathfrak{R}_k, \mathfrak{R}_k] = \mathfrak{R}_k.$$

Thus one has to check only the following remaining cases:

(i) $\mathfrak{R} = M_n(\mathbf{R}) \qquad 1 \le n \le 4$;

(ii) $\mathfrak{R} = M_n(\mathbb{Q}) \qquad n = 1, 2.$

Case (i). For $n = 1$, i.e. $\mathfrak{R} = \mathbf{R}$, the assertion is trivial. If $n \geq 2$ then any skew real martix in an appropriate basis has the form

$$\oplus_{r=1}^{p} \begin{pmatrix} 0 & a_r \\ -a_r & 0 \end{pmatrix} \oplus 0_q,$$

where 0_q is the zero matrix of order q, $2p + q = n$. Since

$$\begin{pmatrix} 0 & a_r \\ -a_r & 0 \end{pmatrix} = \left[\begin{pmatrix} a_r & 0 \\ 0 & 0 \end{pmatrix}, \begin{pmatrix} 0 & 1 \\ 1 & 0 \end{pmatrix} \right]$$

is a commutator of two symmetric matrices, thus any skew element from $M_n(\mathbf{R})$ is also such commutator.

Case (ii). If $n = 1$, i.e. $\mathfrak{R} = \mathbb{Q}$ - quaternions, the assertion is false since $\mathfrak{R}_s = \mathbf{R}$, i.e. $[\mathfrak{R}_s, \mathfrak{R}_s] = \{0\}$, but $\mathfrak{R}_k \neq \{0\}$.

Finally, let $\mathfrak{R} = M_2(\mathbb{Q})$. Then according to Corollary 3.2.11 each skew element is a real linear combination of the following 10 matrices:

$$\begin{pmatrix} i & 0 \\ 0 & i \end{pmatrix}, \begin{pmatrix} j & 0 \\ 0 & j \end{pmatrix}, \begin{pmatrix} k & 0 \\ 0 & k \end{pmatrix}, \begin{pmatrix} i & 0 \\ 0 & -i \end{pmatrix}, \begin{pmatrix} j & 0 \\ 0 & -j \end{pmatrix}$$

$$\begin{pmatrix} k & 0 \\ 0 & -k \end{pmatrix}, \begin{pmatrix} 0 & 1 \\ -1 & 0 \end{pmatrix}, \begin{pmatrix} 0 & i \\ i & 0 \end{pmatrix}, \begin{pmatrix} 0 & j \\ j & 0 \end{pmatrix}, \begin{pmatrix} 0 & k \\ k & 0 \end{pmatrix}$$

But each of the above 10 matrices is a commutator of two symmetric matrices, e.g.

$$\begin{pmatrix} i & 0 \\ 0 & i \end{pmatrix} = \frac{1}{2} \left[\begin{pmatrix} 0 & k \\ -k & 0 \end{pmatrix}, \begin{pmatrix} 0 & j \\ -j & 0 \end{pmatrix} \right],$$

$$\begin{pmatrix} i & 0 \\ 0 & -i \end{pmatrix} = \frac{1}{2} \left[\begin{pmatrix} 0 & i \\ -i & 0 \end{pmatrix}, \begin{pmatrix} 0 & 1 \\ 1 & 0 \end{pmatrix} \right],$$

$$\begin{pmatrix} 0 & 1 \\ -1 & 0 \end{pmatrix} = \frac{1}{2} \left[\begin{pmatrix} 1 & 0 \\ 0 & 0 \end{pmatrix}, \begin{pmatrix} 0 & 1 \\ 1 & 0 \end{pmatrix} \right],$$

$$\begin{pmatrix} 0 & i \\ i & 0 \end{pmatrix} = \frac{1}{2}\left[\begin{pmatrix} 1 & 0 \\ 0 & 0 \end{pmatrix}, \begin{pmatrix} 0 & i \\ -i & 0 \end{pmatrix}\right], \text{ etc.}$$

Therefore in this case $\mathfrak{R}_k = [\mathfrak{R}_s, \mathfrak{R}_s]$ and the proof of Theorem 3.3.7 is complete. ■

Now Theorems 3.3.4 and 3.3.7 imply the following results.

Corollary 3.3.8. *If $\mathfrak{R}$ is a real factor different from $\mathbb{Q}$ and $M_2(\mathbf{R})$, then $[\mathfrak{R}_k, \mathfrak{R}_k] = [\mathfrak{R}_s, \mathfrak{R}_s]$.*

Corollary 3.3.9. *Let $\mathfrak{R}$ be a real factor not of type I_1 or I_2. Then*

1. Any symmetric element from $\mathfrak{R}$ is a sum of a finite number (≤ 25) of squares of skew elements from $\mathfrak{R}$ with coefficients ± 1.

2. $\mathfrak{R} = \overline{[\mathfrak{R}_k, \mathfrak{R}_k]}$.

Proof. Assertion 1 trivially follows from Theorem 3.3.4 since any skew commutator is a skew element.

2. By Corollary 3.3.3, $\mathfrak{R}_s \subset \overline{[\mathfrak{R}_k, \mathfrak{R}_k]}$ and by Theorem 3.3.7 $\mathfrak{R}_k = [\mathfrak{R}_k, \mathfrak{R}_k]$. Therefore

$$\mathfrak{R} = \mathfrak{R}_s + \mathfrak{R}_k \subset \overline{[\mathfrak{R}_k, \mathfrak{R}_k]} + [\mathfrak{R}_k, \mathfrak{R}_k] = \overline{[\mathfrak{R}_k, \mathfrak{R}_k]}. \quad ■$$

Theorem 3.3.10. *Let $\mathfrak{R}$ and S be real factors not of type I_1 and I_2. Then any Lie isomorphism between the Lie algebras $[\mathfrak{R}_k, \mathfrak{R}_k]$ and $[\mathfrak{R}_s, \mathfrak{R}_s]$ can be uniquely extended to an associative *-isomorphism between $\mathfrak{R}$ and S.*

Proof. Let $\mathfrak{R}$ and S be real factors except type I_1 and I_2. By Proposition 3.1.6 they are *-closed algebras over $\mathbf{R}$ with involution of the first kind. We shall consider two cases separately:

1. $\mathfrak{R}$ is of type I_∞, II or III ;

2. $\mathfrak{R}$ is of type I_n, $3 \leq n < \infty$.

Consider the first case. In this case $\mathfrak{R}$ has infinite diminsion and Theorems 3.3.1 and 3.3.4 imply that $[\mathfrak{R}_k, \mathfrak{R}_k]$ is also infinite dimensional. Since $[S_k, S_k]$ is Lie isomorphic to $[\mathfrak{R}_k, \mathfrak{R}_k]$ it follows that $[S_k, S_k]$ and hence S are of infinite dimension, i.e. S also satisfies the condition 1. By Proposition 3.1.7 a real factor is a PI-algebra if and only if it is of type I_n $(n < \infty)$. Therefore neither $\mathfrak{R}$ nor S is PI-algebra over $\mathbf{R}$. Considering the JW-algebra $\mathfrak{R}_s$ of type I_∞, II or III, we can find three mutually orthogonal non zero projection e_1, e_2, e_3 with $e_1 + e_2 + e_3 = \mathbf{I}$ (in the case of type I_∞ we shall choose each e_i to be infinite). Therefore the conditions (a) and (b) of Theorem 3.1.8 are fulfilled. Let us verify the condition (c)

(c) $e \in \overline{M^i}$, where $M^i = [\mathfrak{R}_k, \mathfrak{R}_k] \bigcap \mathfrak{R}^i$, $\mathfrak{R}^i = e_i \mathfrak{R} e_i$, $i = 1, 2, 3$.

Indeed, if $\mathfrak{R}$ is a continuous factor, then applying Corollary 3.3.3 to the continuous factor $\mathfrak{R}^i$ and the projection e^i we obtain

$$e_i \in \mathfrak{R}_i^s \subset \overline{[\mathfrak{R}_i^k, \mathfrak{R}_i^k]} \subset \overline{[\mathfrak{R}^k, \mathfrak{R}^k] \bigcap \mathfrak{R}^i} = \overline{M^i}.$$

Thus all conditions of Theorem 3.1.8 are verified and this theorem implies that the given Lie isomorphism between $[\mathfrak{R}_k, \mathfrak{R}_k]$ and $[S_k, S_k]$ can be uniquely extended to an associative *-isomorphism between $\overline{[\mathfrak{R}_k, \mathfrak{R}_k]}$ and $\overline{[S_k, S_k]}$. By Corollary 3.3.9 (2): $\overline{[\mathfrak{R}_k, \mathfrak{R}_k]} = \mathfrak{R}$ and $\overline{[S_k, S_k]} = S$. ∎

3.4. Isomorphism of Lie operator algebras and conjugacy of involutive anti-automorphisms

Let M be a W^*-algebra, and α an involution (*-anti-automorphism of order 2) on M. Then $\mathfrak{R}^\alpha = \{x \in M : \alpha(x) = x^*\}$ is a real W^*-algebra and $M = \mathfrak{R}^\alpha + i\mathfrak{R}^\alpha$. By Theorem 1.5.10 and Corollary 1.5.11 we obtain (see [S 7, Corollary 3.4])

Theorem 3.4.1. *Let M be a factor with involutions α and β. The following conditions are equivalent:*

(i) $\alpha \sim \beta$;

(ii) the JW-algebras $\mathfrak{R}_s^\alpha$ and $\mathfrak{R}_s^\beta$ are isomorphic;

*(iii) the real factors $\mathfrak{R}^\alpha$ and $\mathfrak{R}^\beta$ are *-isomorphic.*

Let M be a factor with an involution α, $M^\alpha(\pm 1) = \{a \in M : \alpha(a) = \pm a\}$ the spectral subspaces of α. If $a, b \in M^\alpha(+1)$ then $a \circ b, a^* \in M^\alpha(+1)$ therefore $M^\alpha(+1)$ is a self-adjoint Jordan algebra, i.e. a complex Jordan *-subalgebra in M.

Similarly, if $a, b \in M^\alpha(-1)$ then $[a, b]$ and a^* are in $M^\alpha(-1)$, i.e. $M^\alpha(-1)$ is a complex Lie *-subalgebra in M.

Example 3.4.2. Let $M = M_n$ denote the algebra of $n \times n$ complex matrices. Consider the involutions α_r and α_q from the proof of Theorem 1.4.1 :

$$\alpha_r(x) = j_{\mathrm{R}}^{-1} x^* j_{\mathrm{R}}, \quad x \in M_n \quad (\text{if } n \text{ is odd}),$$

and

$$\alpha_q(x) = j_Q^{-1} x^* j_Q, \quad x \in M_n \quad (\text{if } n \text{ even}),$$

where j_{R} is a *conjugation* on H, i.e. a conjugate-linear isometry with $j_{\mathrm{R}}^2 = \mathbf{1}$ and j_Q is an *anti-conjugation* (or a *unit quaternion*), i.e. a conjugate-linear isometry with $j_Q^2 = -\mathbf{1}$. If $n = 2k + 1 (k \geq 1)$ is odd and $\alpha = \alpha_r$ then $M^\alpha(-1)$ coincides with the "classical" Lie algebra B_k. If $n = 2k (k \geq 1)$ is even then for $\alpha = \alpha_r$, $M^\alpha(-1)$ is the Lie algebra D_k and for $\alpha = \alpha_q$ it coincides with C_k (cf. [RS]).

Example 3.4.3. Let $M = B(H)$ be a type I_∞ factor with the canonical involutions α_r and α_q (see Examples 3.4.2 or Theorem 1.4.1). Then the

Lie algebra $M^\alpha(-1)$ for $\alpha = \alpha_r$ is denoted by B_∞ (or D_∞) in [RS] and coincides with the *orthogonal complex Banach Lie algebra*

$$0(H, J_R) = \{x \in B(H): \ J_R x^* J_R = -x\}$$

considered in [Har 1]. For $\alpha = \alpha_q$, $M^\alpha(-1)$ is denoted by C_∞ in [RS] and coincides with the *symplectic complex Banach Lie algebra*

$$sp(H, J_Q) = \{x \in B(H): \ J_Q x^* J_Q = -x\}$$

from [Har 1].

Remark 3.4.4. Consider a factor M with an involution α. It is clear that $\mathfrak{R}_s^\alpha = M^\alpha(+1)_s$ and $\mathfrak{R}_k^\alpha = M^\alpha(-1)_k$. Therefore the study of Jordan and Lie algebras $M^\alpha(+1)$ and $M^\alpha(-1)$ can be reduced to the study of self-adjoint and skew-adjoint elements in real factors.

Now let α and β be involutions of a factor M. From Theorem 3.4.1 and Remark 3.4.4 it follows that the Jordan algebras $M^\alpha(+1)$ and $M^\beta(+1)$ are complex *-isomorphic if and only if α and β are conjugate.

We shall prove now the main result of the chapter.

Theorem 3.4.5. *Let M be a factor, α and β involutions on M. The Lie algebras $M^\alpha(-1)$ and $M^\beta(-1)$ are isomorphic, as complex Lie *-algebras, if, and only if, α and β are conjugate.*

Proof. If α and β are conjugate then by Theorem 3.4.1 the real factor $\mathfrak{R}^\alpha$ is *-isomorphic to the real factor $\mathfrak{R}^\beta$ and therefore $\mathfrak{R}_k^\alpha$ and $\mathfrak{R}_k^\beta$ are isomorphic as Lie algebras. Thus $M^\alpha(-1)$ and $M^\beta(-1)$ are complex *-isomorphic Lie algebras.

Conversely, suppose that Lie algebras $M^\alpha(-1)$ and $M^\beta(-1)$ are isomorphic. If M is a factor not of type I_1 or I_2 then $\mathfrak{R}^\alpha$ and $\mathfrak{R}^\beta$ are real factors of the same type, such that their skew element $\mathfrak{R}_k^\alpha$ and $\mathfrak{R}_k^\alpha$ are isomorphic as Lie algebras. By Theorem 3.3.10 $\mathfrak{R}^\alpha$ and $\mathfrak{R}^\beta$ are

*-isomorphic and by Theorem 3.4.1 involutions α and β are conjugate.

Thus it remains to coinsider the type I_1 or I_2. For sake of completeness we shall consider arbitrary type I_n $(n < \infty)$ factors.

Let $M = B(H)$ be a type I_n factor. If $dim H = n = 2m + 1$ is odd (in particular if M is of type I_1) then there is a unique (up to conjugacy) involution $\alpha = \alpha_r$ and $M^\alpha(-1) = B_m$ (see Examples 4.3.2 and 4.3.3) so the assertion is trivial. If $dim H = n = 2m$ is even (in particular if M is of type I_2) then there are exactly two involutions $\alpha = \alpha_r$ and $\beta = \alpha_q$ and the corresponding Lie algebras $M^\alpha(-1) = D_m$ and $M^\beta(-1) = C_m$ are known to be non isomorphic (see Examples 4.3.2 and 4.3.3) since $dim D_m = 2m^2 - m$, $dim C_m = 2m^2 + m$. Therefore the proof of Theorem 3.4.5 is complete. ■

COMMENTS TO CHAPTER 3

3.1. For the definition of the extended centroid we followed the exposition of [CR]. Algebraic properties of real factors have been considered in [A 17]. The problem of extension of Lie isomorphisms between Lie algebras $[\mathfrak{R}_k, \mathfrak{R}_k]$ and $[S_k, S_k]$ to an associative *-isomorphism from $\mathfrak{R}$ to S for the case of simple algebras (rings) was considered by Martindale in [Ma 1] for the case of prime rings with involution of the first kind - in [BMM], [Ma 2], for prime rings with an arbitrary involution in [Roz]. In exposition of Theorem 3.1.8 we followed [Roz].

3.2. Representations of symmetric elements of a simple ring with involution as a sum of squares of skew-symmetric elements with coefficients ± 1 were obtained by Baxter [Bax]. Our Theorems 3.2.1 and 3.2.7 extend this result to non simple rings $B(H)$ and moreover give an estimate for the number of terms in the sum, where H is a real, complex or quaternionian Hilbert space.

Further results are concerned with a problem suggested by Herstein in his paper [He 1]: given a simple ring $\mathfrak{R}$ with involution is it true that $\mathfrak{R}_k = [\mathfrak{R}_k, \mathfrak{R}_k]$? A negative answer was given by Lee in [L], who has constructed an example of a division ring with involution in which some skew elements cannot be represented as commutators.

For algebras $B(H)$, where H is real or quaternionian Hilbert space, a positive answer to the Herstein problem was given by Sadovskij [Sad], who has proved that if $dim H \geq 3$, then any skew elements is a commutator of two skew elements.

3.3. The results of this Section generalise the results of Section 3.2

from real factors of type I to arbitrary real factors. On the other hand they can be interpreted as an extension of well-known results about algebraically simple algebras to topologically simple algebras, since W^*-factors are exactly von Neumann algebras without non zero proper weakly closed (two-sided) ideals.

The main Theorem 3.3.10 also generalizes to real factors a theorem of Martindale [Ma 1] on Lie isomorphisms of skew elements in simple rings.

Besides, the results of Sections 3.2 and 3.3 give an additional deep information on properties of single operators in Hilbert spaces. These results were obtained in papers of Ayupov [A 16], [A 18], Ayupov and Azamov [AAz].

3.4. Pairs of Lie and Jordan algebras of the form $M^\alpha(\pm 1)$ were firstly considered in the work of Robinson and Stormer [RS] for the case $M = B(H)$ and properties of Lie algebras $M^\alpha(-1)$ were also studied in [Har 1].

The main Theorem 3.4.5 for type I factors was proved by Ayupov in [A 15] and for type II_1 and σ-finite type III factors in [A 17], later for arbitrary factors but in a weaker form in [A 16],[A 18]. The final version of the theorem was obtained by Ayupov and Azamov [AAz].

References

[ASS] Alfsen, E., Shultz, F. and Stormer, E. A Gelfand-Neumark theorem for Jordan algebras
Advances in Math., 1978. Vol. 28. N 1. pp. 11-56

[Ar] Arveson W. On groups of automorphism of operator algebras
Journal of Functional Analysis, 1974. N 15. pp. 217-243

[A 1] Ayupov Sh.A. Extension of traces and type criterions for Jordan algebras of self-adjoint operators
Mathematische Zeitscrift, 1982. Vol. 181. pp. 253-268

[A 2] Ayupov Sh.A. Modular Jordan algebras of self-adjoint operators
Theor. i mat. phyzika. 1982. Vol. 53. N 1. pp. 77-82 (russian).

[A 3] Ayupov Sh.A. On the construction of Jordan algebras of self-adjoint operators
Dokl. Acad. Nauk SSSR. 1982. Vol. 267. N 3. pp. 521-524 (russian). English transl. in *Soviet Math. Dokl.* Vol. 26, (1982) N 3. pp. 623-626

[A 4] Ayupov Sh.A. Types of Jordan algebras of self-adjoint operators and their enveloping von Neumann algebras
Funkt. Anal. i Ego Pril. 1983. Vol. 17. N 1. pp. 65-66 (russian).

[A 5] Ayupov Sh.A. Classification of injective JW-factors
Funkt. Anal. i Ego Pril. 1984. Vol. 18. N 3. pp. 67-68 (russian).

[A 6] Ayupov Sh.A. On existence of Jordan algebras of self-adjoint operators of a given type
Sibir. Mat. J., 1984. Vol. 25. N 5. pp. 3-8 (russian). English transl. in *Siberian mathematical J.* Vol. 25. N 1-6. (1984), pp. 689-693

[A 7] Ayupov Sh.A. *JW*-factors and anti-automorphisms of von Neumann algebras
Izv. Acad. Nauk SSSR. Ser. Mat. 1985. Vol. 49. N 1. pp. 211-220 (russian); English transl. in *Math. USSR - Izvestiya* 1986. Vol. 26. pp. 201-209

[A 8] Ayupov Sh.A. Jordan operator algebras
Sovrem. Prob. Mat., Itogi Nauki i Tekhniki VINITI. Moscow, 1985. Vol. 27. pp. 67-98 (russian).

[A 9] Ayupov Sh.A. Classification and Representation of Ordered Jordan algebras
1986. "Fan". Tashkent. 121 p. (russian).

[A 10] Ayupov Sh.A. Traces on *JW*-algebras and enveloping W^*-algebras
Mathematische Zeitscrift, 1987. Vol. 194. pp. 15-23

[A 11] Ayupov Sh.A. Diameters of state spaces of Jordan Banach algebras
Baku International Topological Conference. Abstracts p. 11. 1987. 22 p.

[A 12] Ayupov Sh.A. Traces on *JW*-algebras and enveloping von Neumann algebras
Collection: *Operatornie Algebri i Funktionalnie prostranstwa.* Tashkent. "Fan". 1988. pp. 3-11 (russian).

[A 13] Ayupov Sh.A. A new proof of the existence of traces on Jordan operator algebras and real von Neumann algebras
Journal of Functional Analysis, 1989. Vol. 84. N 2. pp. 312-321

[A 14] Ayupov Sh.A. On existence of traces on modular *JW*-factor
Uspehi Mat. Nauk. 1989, Vol. 44. N 1. (265), pp. 183-184 (russian).

[A 15] Ayupov Sh.A. Anti-automorphisms and Lie structure in factors
Uzbek. Mat. J., 1992. N 2. pp. 8-13 (russian).

[A 16] Ayupov Sh.A. Skew commutators and Lie isomorphisms in real von Neumann algebras
Preprint 1994 / 012 IRMA Strasbourg University, 20 p.

[A 17] Ayupov Sh.A. Anti-automorphisms of factors and Lie operator algebras
Quart. J. Math. Oxford ser, 1995. Vol. 46. N 2. pp. 129-140

[A 18] Ayupov Sh.A. Skew commutators and Lie isomorphisms in real von Neumann algebras

Journal of Functional Analysis, 1996. Vol. 138. N 1. pp. 170-187

[AAz] Ayupov Sh.A., Azamov N.A. Commutators and Lie isomorphisms of skew elements in prime operator algebras
Communications in Algebra, 1996. 24(4). pp. 1501-1520

[AAd] Ayupov Sh.A., Adizov A.A. Probability measures on projections of JBW-algebras
Preprint 7822-84 *deposed at* VINITI 1984. (russian).

[AU 1] Ayupov Sh.A., Usmanov Sh.M. Diameters of state spaces of Jordan Banach algebras
Conference: *Topologicheskaya Algebra*. 1988. Kishinew, Shtiinza. 10 p. (russian).

[AU 2] Ayupov Sh.A., Usmanov Sh.M. Diameters of state spaces of Jordan Banach algebras
Izv. Acad. Nauk SSSR. Ser. Mat. 1089. Vol. 53. N 2. pp. 227-242 (russian). English transl. in *Math. USSR Izvestiya*, Vol. 34. (1990), N 2, pp. 229-244

[Bax] Baxter W. Lie simplicity of a special class of associative rings. II
Trans. Amer. Math. Soc. 1958. Vol 87. pp. 63-75

[BMM] Beidar K., Martindale W., Mikhalev A. Lie isomorphisms in prime rings with invilution
J. of Algebra, 1994. Vol 169. pp. 304-327

[Boy 1] Boykabilov B.M. Injective real von Neumann algebras
Preprint 1014-B94 *deposed at* VINITI 1994. 26 p. (russian).

[Boy 2] Boykabilov B.M. Injective real von Neumann factors
Dokl. Acad. Nauk Uzbek. 1995. N 7-8. pp. 3-4 (russian).

[BU] Boykabilov B.M., Usmanov Sh.M. Completely positive maps on real factor of type III_λ, $0 < \lambda < 1$
Dokl. Acad. Nauk Uzbek. 1996. N 4. pp. 6-8 (russian).

[BR] Bratteli O., Robinson D. Operator algebras and quantum statistical mechanics
Springer-Verlag, New York Heidelber, Berlin 1979.

[BW] Bunce L.J., Wright J.D.M. Quantum measures and states on Jordan algebras
Commun. Math. Phys. 1985. Vol 98. pp. 187-202

[CR] Cabrera M., Rodrigues A. Extended centroid and central closure of semiprime normed algebras A first approach,
Commun. Algebra 1990. Vol 18. pp. 2293-2326

[CL] Chuang C.L., Lee P.H. Skew commutators and simple rings *J. of Algebra* 1982. Vol 74. pp. 280-281

[C 1] Connes A. Une classification des facteurs de type III *Ann. Sc. Ec. Norm. Sup.*, 1973, 6. pp. 133-252

[C 2] Connes A. Outer conjugacy classes of automorphisms of factors *Ann. Scient. Ec. Norm. Sup.*, 4^e serie, t. 8 1975. pp. 383-420

[C 3] Connes A. Classification of automorphisms of hyperfinite factors of type II_1 and II_∞ and application to type III factors *Bull. Amer. Math. Soc.* 1975. Vol. 81. N 6. pp. 1090-1092

[C 4] Connes A. Classification of injective factors *Ann. Math.*, 1976. Vol. 104. N 1. pp. 73-115

[C 5] Connes A. Periodic automorphisms of the hyperfinite factor of type II_1 *Acta Sci. Math.* 1977. N 39. pp. 39-66

[CHS] Connes A., Haagerup U., Stormer E. Diameters of state spaces of type III factors *Operator Algebras and Their Connections with Topology and Ergodic Theory* (Busteni, 1983), Lecture Notes in Math. 1985. Vol. 1132, Springer-Verlag, pp. 91-116

[CS] Connes A., Stormer E. Homogeneity of the state space of factors of type III_1 *Journal of Functional Analysis*, 1978. Vol. 28. pp. 91-116

[CT 1] Connes A., Takesaki M. Flots de poids sur les facteurs de type III *C. R. Acad. Sci. Paris. Ser. A.*, 1974. T. 278. pp. 937-940

[CT 2] Connes A., Takesaki M. The flow of weights on factors of type III *Thoku Math. J. Ser.* 1977. Vol. 29. N 2. pp. 473-575

[CM] Mc Crimmon K., Montgomery S. Open questions on rings with involution in Ring Theory Conference, University of Chicago, July 1973.

[Da] Van Daele A. Continuous crossed products and type III von Neumann algebras *London Math. Soc.*, Lecture Notes 31 (Cambridge University Press, London, 1978).

[Dix 1] Dixmier J. Les algebres d'operateurs dans l'espace Hilbertien Paris: Gauthier-Villars. 1969. 369 p.

[Dix 2] Dixmier J. Les C^*-algebres et leurs representations. Paris, Gauthier - Villars ed. 1969.

[Dix 3] Dixmier J. Von Neumann algebras, Universite Paris. 6., France, 1981.

[D 1] McDuff D. To the structure of II_1-factors *Uspehi Mat. Nauk*. 1970. Vol. XXV. N 6. 156. pp. 29-53 (russian).

[D 2] McDuff D. Central sequences and the hyperfinite factor *Proc. London Math. Soc.* 1970. Vol 21. pp. 443-461

[EL] Effros E., Lance E. Tensor products of operator algebras *Advances in Math.* 1977. Vol 25. 34 p.

[ES] Effros E., Stormer E. Jordan algebras of self-adjoint operators *Trans. Amer. Math. Soc.*, 1967. Vol. 127. N 2. pp. 313-316

[EW] Elliott G.A., Woods E.J. The equivalence of various definition for a properly infinite von Neumann algebra to be approximately finite dimensional *Proc. Amer. Math. Soc.* 1976. Vol 60. pp. 175-179

[E] Emch G. Algebraic methods in statistical mechanics and quantum fields theory John Viley and Sons, Inc. London - New - York - Toronto. 1972.

[FH] Fack Th., de la Harpe P. Sommes de commutateurs dans les algebres de von Neumann finies continues *Ann. Inst. Fourier*, Grenoble 30. 1980., pp. 49-73.

[GJ] Giordano T., Jones V. Antiautomorphismes involutifs du facteur hyperfini de type II_1 *C. R. Acad. Sci. Paris. Ser. A*, 1980. T. 290. pp. 29-31

[G 1] Giordano T. Antiautomorphismes involutifs des facteurs injectifs *C. R. Acad. Sci. Paris Ser. A.* 1980. T 291. pp. 583-585

[G 2] Giordano T. Antiautomorphismes involutifs des facteurs de von Neumann injectifs These , Universite de Neuchatel, 1981. 106 p.

[G 3] Giordano T. Antiautomorphismes involutifs des facteurs de von Neumann injectifs. II *Journal of Functional Analysis.* 1983. Vol. 51. N 3. pp. 326-360

[G 4] Giordano T. Antiautomorphismes involutifs des facteurs de von Neumann injectifs. I *J. Operator theory*, 1983. Vol. 10. N. 2 pp. 252-287

[G 5] Giordano T. Classification of antiautomorphisms of injective factors of type III_λ, $\lambda \neq 0$
Current topics in operator algebras (Nara, 1990). pp. 425-431, World Sci. Publishing, River Edge,, NJ, 1991.

[H 1] Haagerup U. Connes' bicentralizer problem and uniqueness of the injective factor of type III_1
Preprint. N 10. Odense University. 1984. 68 p.

[H 2] Haagerup U. A new proof of the equivalence of injectivity and hyperfiniteness for factors on a separable Hilbert space
Journal of Functional Analysis. 1985. Vol. 62. pp. 160-201

[HS] Haagerup U., Stormer E. Positive projections of von Neumann algebras onto JW-algebras
J. Reports on Math. Physics. 1995. Vol. 36. pp. 317-330

[Ha 1] Halmos P.R. Commutators of operators, II
Amer. J. Math. 1954. Vol. 76. pp. 191-198

[Ha 2] Halmos P.R. A Hilbert space problem book
D. van Nostrand, Princeton, New Jersey, Toronto London, 1967.

[H-OS] Hanche-Olsen H., Stormer E. Jordan operator algebras, London: Pitman Ltd, 1984. VIII+183 p.

[Har 1] de la Harpe P. Classical Banach-Lie algebras and Banach-Lie groups of Operators in Hilbert spaces
Lect. Notes Math. 285. 1972., Springer-Verlag, Berlin /Heidelberg/ New-York.

[Har 2] de la Harpe P. Classical groups and classical Lie algebras of operators
Proc. Symp. Pure Math. AMS 38-1, 1982. pp. 477-513

[He 1] Herstein I.N. Lie and Jordan structure in simple associative rings
Bull. Amer. Soc. 67. 1961. pp. 517-531

[He 2] Herstein I.N. Topics in Ring Theory
University of Chicago press. IL, 1969.

[He 3] Herstein I.N. Non commutative rings
"Mir". Moscow. 1972. (russian Edition).

[Ja] Jacobson N. Structure and Representations of Jordan Algebras
Amer. Math. Soc. Colloq. Publ. 39, 1968. X+453 p.

[JSSS] Jevlakov K., Slin'ko A., Shestakov I., Shirshov A. Rings near to associative
"Nauka". Moscow. 1978. 432 p. (russian).

[JN] Jordan P., von Neumann J. Wigner E. On an algebraic generalization of the quantum mechanical formalism
Ann. Math. 1934. Vol. 35. pp. 29-64

[K 1] Kaplansky I. Normed algebras
Duke Math. J. 1949. Vol. 16. pp. 399-418

[K 2] Kaplansky I. Any orthocomplemented complete lattice is a continuous geometry
Ann. Math. 1955. Vol. 16. pp. 524-547

[KG] Kirillov A.A., Gvishiany A.D. Theorems and problems in Functional Analysis
"Nauka". Moscow. 1979. (russian).

[Kr] Krieger W. On non singular transformation of a measure space
I.Z.Wahrscheinlickkeitstheorie werw. - Geb. Bd. 17. 1969, pp. 83-97

[L] Lee P.H. An example of divisions rings with involution
J. of Algebra, 74. 1982. pp. 282-283

[Ma 1] Martindale,3rd, W.S. Lie isomorphisms of the skew elements of a simple ring with involution
J. of Algebra, 36. 1975. pp. 408-415

[Ma 2] Martindale,3rd, W.S. Lie isomorphisms of the skew elements of a simple ring with involution
Commun. Algebra, 4. 1976. pp. 929-977

[M] Mathieu M. Elementary operators on prime C^*-algebras
I. Math. Ann. 284. 1989. pp. 223-244

[MN 1] Murray F., von Neumann J. On rings of operators. I
Ann. Math., 1936. Vol. 37. pp. 116-229

[MN 2] Murray F., von Neumann J. On rings of operators. II
Trans. Amer. Math. Soc., 1937. Vol. 41. pp. 208-248

[MN 3] Murray F., von Neumann J. On rings of operators. IV
Ann. Math. 1943. Vol. 44. pp. 716-808

[Ne] Neimark M.A. Normed rings
"Nauka". Moscow. 1968. 664 p.

[N 1] von Neumann J. Continuous geometry and other topics
Collected works. 1962. Vol. IV. Oxford: Pergamon Press, X + 516p.

[N 2] von Neumann J. Selected works in functional analysis
"Nauka". Moscow. 1987. Vol. 1. 376 p. (russian).

[N 3] von Neumann J. Selected works in functional analysis
"Nauka". Moscow. 1987. Vol. 2. 370 p. (russian).

[PT] Pearcy C., Topping D. Sums of small numbers of idempotents
Michigan Math. J. 14. 1968. pp. 453-465

[Pe] Pedersen G.K. C^*-algebras and their automorphism groups
Academic Press, London, New York, San Fransisco, 1979.

[PeT] Pedersen G.K., Takesaki M. The Radon-Nikodym Theorem for von Neumann algebras
Acta Math. 130. 1973. pp. 53-87

[PeS] Pedersen G., Stormer E. Traces on Jordan algebras
Canad. J. Math. 1982. Vol. 34. N 2. pp. 370-373

[Pon] Pontrjagin L.S. Continuous Groups
Nauka, Moscow 1973. (russian).

[P] Popa S. A short proof of "injectivity implies hyperfiniteness" for finite von Neumann algebras
J. Operator Theory, 16. 1986. pp. 261-272

[Pow] Powers R.T. Representation of uniformly hyperfinite algebras and their associated von Neumann rings
Ann. Math. 1967. Vol. 86. N 1. pp. 138-171

[R 1] Rakhimov A.A. Classification of periodic *-antiautomorphisms of type II_1 hyperfinite factor
Dokl. Acad. Nauk. Uz.SSR. 1991. N 10. pp. 3-4 (russian).

[R 2] Rakhimov A.A. Periodic automophisms and antiautomophisms of injective factors of type II_1
Dissertation, Tashkent. Institute of Mathematics of Uzbek Academy of Sci., 1992. 126 p. (russian).

[R 3] Rakhimov A.A. To the structure real factors of type III_λ, $0 < \lambda < 1$
Collection Tashkent State University *Nekotorie voprosi Analisa i Algebri.* 1995. N 1. pp. 77-98 (russian).

[R 4] Rakhimov A.A. Non-commutative real analogue of Rokhlin's lemma *Dokl. Acad. Nauk. Uzbek.* 1995. N 4. pp. 12-13 (russian).

[R 5] Rakhimov A.A. Injective and hyperfinite real W^*-factors of type III_λ, $0 < \lambda < 1$ *Dokl. Acad. Nauk Uzbek.*, 1997, N1, pp. 3-5 (russian).

[R 6] Rakhimov A.A. Injective real W^*-factors of type III_λ, $0 < \lambda < 1$ *Funkt. Anal. i Ego Pril.* (to paper) (russian).

[RU 1] Rakhimov A.A., Usmanov Sh.M. Classification of periodic anti-automorphisms with odd inner semi-period of the hyperfinite factor of type II_1 *Preprint* 7293-B88 *deposed at* VINITI 1988. 48 p. (russian).

[RU 2] Rakhimov A.A., Usmanov Sh.M. Classification of periodic anti-automorphisms with odd inner semi-period of the hyperfinite factor of type II_1 *Dokl. Acad. Nauk UzSSR.* 1989. N 8. pp. 3-4 (russian).

[RU 3] Rakhimov A.A., Usmanov Sh.M. Classification of periodic *-automorphisms and *-anti-automorphisms of type II_1 hyperfinite real factor *Dokl. Acad. Nauk UzSSR.* 1991. N 11. pp. 5-6 (russian).

[RU 4] Rakhimov A.A., Usmanov Sh.M. Periodic *-automorphisms and *-anti--automorphisms of inlective factors *Funkt. Anal. i Ego Pril.* 1995. Vol. 29. N 1. pp. 87-89 (russian). English transl. 1995. N 1. pp. 68-70

[RU 5] Rakhimov A.A., Usmanov Sh.M. Outer conjugacy classes of automorphisms of type II_∞ factors and a real factors of type III_λ $(0 < \lambda < 1)$ *Dokl. Akad. Nauk Uzbek.* 1996. N 10. pp. 3-5 (russian).

[RU 6] Rakhimov A.A., Usmanov Sh.M. Outer conjugacy classes of automorphisms and antiautomorphisms of real and complex injective factors *Journal of Functional Analysis*, 1997. Vol. 144.

[RS] Reed M., Simon B. Methods of Modern Mathematical Physics 1. Functional Analysis. Academic Press, New York, London, 1972.

[Ro] Robertson A.G. Automorphisms of spin factors and the decomposition of positive maps *Quart. J. Math. Oxford*, 1983. Vol. 34. N 133. pp. 87-96

[RS] Robinson D., Stormer E. Lie and Jordan structure in operator algebras *J. Austral. Math. Soc.*, 1980. Vol. A29. N 2. pp. 129-142

[Roz] Rozen M.P. Isomorphisms of a certain class of prime Lie rings
J. of Algebra, 89. 1984. pp. 291-317

[Sad] Sadovskij V.Yu. On skew-hermitian elements of real W^*-factors of type I
Uzbek Math. J. 1994. N 1. pp. 55-63 (russian).

[Sak 1] Sakai S. C^*-algebras and W^*-algebras
Berlin: Springer, 1971. IX+256 p.

[Sak 2] Sakai S. On automorphism groups of II_1-factors
Tohoku Math. J. 1974. Vol. 26. N 3. pp. 423-430

[SAKC] Sarymsakov T., Ayupov Sh., Khadjiev Dj., Chilin V. Ordered Algebras
"Fan". Tashkent. 1983. 304 p. (russian).

[Sh] Shultz F.W. On normed Jordan algebras which are Banach dual spaces
Journal of Functional Analysis. 1979. Vol. 31. N 3. pp. 360-376

[St 1] Stacey P.J. Real structure in the approximately finite dimensional II_∞ factor
Preprint Melbourne, 1981.

[St 2] Stacey P.J. The structure of type I JBW-algebras
Math.Proc. Camb. Phil. Soc. 1981. Vol. 90. N 3. pp. 477-482

[St 3] Stacey P.J. Type I_2 JBW-algebras
Quart. J. Math. Oxford, 1982. Vol. 33 N 129 pp. 115-127

[St 4] Stacey P.J. Real structure in σ - finite factors of type III_λ where $0 < \lambda < 1$
Proc. London Math. Soc. 1983. (3) 47. pp. 275-284

[S 1] Stormer E. On the Jordan structure of C^*-algebras
Trans. Amer. Math. Soc. 1965. Vol. 120. N 12. pp. 438-447

[S 2] Stormer E. Jordan algebras of type I
Acta Math., 1966. Vol. 115. N 3-4. pp. 165-184

[S 3] Stormer E. On anti-automorphisms of von Neumann algebras
Pacific J. Math. 1967. Vol. 21. N. 2. pp. 349-370

[S 4] Stormer E. Irreducible Jordan algebras of self-adjoint operators
Trans. Am. Math. Soc. 1968. 130. pp. 153-166

[S 5] Stormer E. Real structure in the hyperfinite factor
Duke Math. J. 1980. Vol. 47. N 1. pp. 145-153

[S 6] Stormer E. Conjugacy of involutive antiautomorpisms of von Neumann algebras
Preprint Oslo University, 1984. N 4. 20 p.

[S 7] Stormer E. Conjugacy of involutive antiautomorpisms of von Neumann algebras
Journal of Functional Analysis. 1986. Vol. 66. pp. 54-66

[Str] Stratila S. Modular theory in operator algebras
Bucuresti: Editura Academiei; Tunbridge Wells: Abacus Press. 1981. 492 p.

[SZ] Stratila S., Zsido L. Lectures on von Neumann algebras
Bucuresti: Editura Academiei; Tunbridge Wells: Abacus Press. 1979. 478 p.

[Su] Sunouchi H. Infinite Lie rings
Tohoku Math. J. 1980. Vol. 47. pp. 145-153

[T 1] Takesaki M. Theory of operator algebras, I
Berlin: Springer. 1979. VIII + 415 p.

[T 2] Takesaki M. The structure of a von Neumann algebra with a homogeneous periodic state
Acta math. 131. 1973. pp. 79-122

[T 3] Takesaki M. Conditional expectation in von Neumann algebra
Journal of Functional Analysis. 1972. Vol. 9. pp. 306-321

[T 4] Takesaki M. Structure of factors and automorphism groups
Amer. Math. Soc. 1983. 107 p.

[To 1] Topping D. Jordan algebras of self-adjoint operators
Mem. Amer. Math. Soc. 53. Providence R. I.: AMS. 1965. 48 p.

[To 2] Topping D. An isomorphism invariant for spin factors
J. Math. Mech. 1966. Vol. 15. N. 6. pp. 1055-1064

[U 1] Usmanov Sh.M. Classification of real W^*-factors of type $III_\lambda, 0 < \lambda < 1$
Preprint 8082-84 *deposed at* VINITI 1984. (russian).

[U 2] Usmanov Sh.M. Structure of real W^*-factors of type III_0
Dokl. Acad. Nauk Uz.SSR. 1984. N 9. pp. 3-4 (russian).

[U 3] Usmanov Sh.M. Classification of real W^*-factors of type $III_\lambda, 0 \leq \lambda < 1$
Funkt. Anal. i Ego Pril. 1985. Vol. 19. N 3. pp. 94-95

[U 4] Usmanov Sh.M. Structure and Classification of Jordan algebras of self-adjoint operators
Dissertation, Tashkent State Univ. 1985. 137 p. (russian).

[Up] Upmeier H. Automorphism groups of Jordan C^*-algebras
Mathematische Zeitscrift. 1981. Vol. 176. pp. 21-34

[W] Wasserman S. Injective W^*-algebras
Math. Proc. Cambridge Phil. Soc. 82. 1977. pp. 39-47

INDEX

Algebra
  central sequences of 113
  JW-algebra 14
    abelian 18
    injective 67
    locally modular 23
    modular 23
    reversible 14
    properly non-modular 23
    purely real 15
    purely non-modular 23
  Lie 174
  PI (PI-algebra) 180
  prime 175
  real (W^*-algebra) 15
    amenable 145
    finite 24
    hyperfinite 147
    injective 67, 116
    properly infinite 24
    purely infinite 24
    semidiscrete 142
    semifinite 24
  semiprime 175
  simple 175
  von Neumann (W^*-algebra) 7
    infinite 8
    hyperfinite 58, 120
    discrete 8
    injective 58
    finite 8
    continuous 8
    semifinite 8
    properly infinite 8
    purely infinite 8
    σ-finite 8
  *-closed prime 176
*-automorphism 9
  aperiodic 135
  canonical purely periodic 121
  centrally trivial 135
  inner 9
  minimally periodic 111
  periodic 110
  properly outer 130
  purely periodic 110
*-antiautomorphism 9
  aperiodic 135
  canonical purely periodic 120
  involutive 49
  periodic 110
  properly outer 130
  purely periodic 110
Asymptotical period
  of *-antiautomorphisms 135
  of *-automorphism 135
Center of W^*-algebra 7
Centroid of algebra 176
Conjugacy
  of antiautomorphisms 9, 59
  of automorphisms 9, 59
Crossed product 52, 75
Diameter of state spaces
  of JW-algebra 152
  of W^*-algebra 152
Element self-adjoint 181

skew-adjoint 181
Enveloping W^*-algebra 15
Equivalence of projections 8, 24
Essentially defined centralizer 175
Extended centroid 176
Factor 7
Mc Duff 59
Krieger 52, 54
Flow of weights 60, 61
Fundamental homomorphism 61
Involution of the first kind 176
of the second kind 176
Lattices of projections
of JW-algebra 23
modular 23
Maximal partial defined centralizer 176
Modulo
of *-antiautomorphism 61
of *-automorphism 61
Modular automorphism groups 11
Modular involution 11
Morphism 141
completely positive 141
n-positive 141
Partially defined centralizer 175
Projection
abelian 8
finite 8
infinite 8
purely infinite 8
Spectrum of W^*-algebra 82
Spin-factor 15
State 9
faithful 10
normal 10
semifinite 10
Trace
on W^*-algebra 10
on JW-algebra 30
faithful 30
finite 30
normal 30
semifinite 30
Type I (I_n, I_n, I_∞)
JW-algebra 38
W^*-algebra 8
real W^*-algebra 24
W^*-factor 8
real W^*-factor 24
II (II_1, II_∞)
JW-algebra 38
W^*-algebra 8, 9
real W^*-algebra 24
W^*-фактор 8
real W^*-factor 24
III (III_1, III_0, III_λ)
JW-algebra 38, 52, 67
W^*-algebra 9, 12
real W^*-algebra 24, 67
W^*-factor 9, 12
real W^*-factor 24, 67
Weight 9
faithful 10
finite 10
infinite multiplicity 94
lacunary 94
normal 10
semifinite 10
strictly semifinite 78

Other *Mathematics and Its Applications* titles of interest:

P.H. Sellers: *Combinatorial Complexes. A Mathematical Theory of Algorithms.* 1979, 200 pp. ISBN 90-277-1000-7

P.M. Cohn: *Universal Algebra.* 1981, 432 pp.
ISBN 90-277-1213- 1 (hb), ISBN 90-277-1254-9 (pb,

J. Mockor: *Groups of Divisibility.* 1983, 192 pp. ISBN 90-277-1539-4

A. Wwarynczyk: *Group Representations and Special Functions.* 1986, 704 pp.
ISBN 90-277-2294-3 (pb), ISBN 90-277-1269-7 (hb)

I. Bucur: *Selected Topics in Algebra and its Interrelations with Logic, Number Theory and Algebraic Geometry.* 1984, 416 pp. ISBN 90-277-1671-4

H. Walther: *Ten Applications of Graph Theory.* 1985, 264 pp.
ISBN 90-277-1599-8

L. Beran: *Orthomodular Lattices. Algebraic Approach.* 1985, 416 pp.
ISBN 90-277-1715-X

A. Pazman: *Foundations of Optimum Experimental Design.* 1986, 248 pp.
ISBN 90-277-1865-2

K. Wagner and G. Wechsung: *Computational Complexity.* 1986, 552 pp.
ISBN 90-277-2146-7

A.N. Philippou, G.E. Bergum and A.F. Horodam (eds.): *Fibonacci Numbers and Their Applications.* 1986, 328 pp. ISBN 90-277-2234-X

C. Nastasescu and F. van Oystaeyen: *Dimensions of Ring Theory.* 1987, 372 pp.
ISBN 90-277-2461-X

Shang-Ching Chou: *Mechanical Geometry Theorem Proving.* 1987, 376 pp.
ISBN 90-277-2650-7

D. Przeworska-Rolewicz: *Algebraic Analysis.* 1988, 640 pp. ISBN 90-277-2443-1

C.T.J. Dodson: *Categories, Bundles and Spacetime Topology.* 1988, 264 pp.
ISBN 90-277-2771-6

V.D. Goppa: *Geometry and Codes.* 1988, 168 pp. ISBN 90-277-2776-7

A.A. Markov and N.M. Nagorny: *The Theory of Algorithms.* 1988, 396 pp.
ISBN 90-277-2773-2

E. Kratzel: *Lattice Points.* 1989, 322 pp. ISBN 90-277-2733-3

A.M.W. Glass and W.Ch. Holland (eds.): *Lattice-Ordered Groups. Advances and Techniques.* 1989, 400 pp. ISBN 0-7923-0116-1

N.E. Hurt: *Phase Retrieval and Zero Crossings: Mathematical Methods in Image Reconstruction.* 1989, 320 pp. ISBN 0-7923-0210-9

Du Dingzhu and Hu Guoding (eds.): *Combinatorics, Computing and Complexity.* 1989, 248 pp. ISBN 0-7923-0308-3

Other *Mathematics and Its Applications* titles of interest:

A.Ya. Helemskii: *The Homology of Banach and Topological Algebras*. 1989, 356 pp. ISBN 0-7923-0217-6

J. Martinez (ed.): *Ordered Algebraic Structures*. 1989, 304 pp. ISBN 0-7923-0489-6

V.I. Varshavsky: *Self-Timed Control of Concurrent Processes. The Design of Aperiodic Logical Circuits in Computers and Discrete Systems*. 1989, 428 pp. ISBN 0-7923-0525-6

E. Goles and S. Martinez: *Neural and Automata Networks. Dynamical Behavior and Applications*. 1990, 264 pp. ISBN 0-7923-0632-5

A. Crumeyrolle: *Orthogonal and Symplectic Clifford Algebras. Spinor Structures*. 1990, 364 pp. ISBN 0-7923-0541-8

S. Albeverio, Ph. Blanchard and D. Testard (eds.): *Stochastics, Algebra and Analysis in Classical and Quantum Dynamics*. 1990, 264 pp. ISBN 0-7923-0637-6

G. Karpilovsky: *Symmetric and G-Algebras. With Applications to Group Representations*. 1990, 384 pp. ISBN 0-7923-0761-5

J. Bosak: *Decomposition of Graphs*. 1990, 268 pp. ISBN 0-7923-0747-X

J. Adamek and V. Trnkova: *Automata and Algebras in Categories*. 1990, 488 pp. ISBN 0-7923-0010-6

A.B. Venkov: *Spectral Theory of Automorphic Functions and Its Applications*. 1991, 280 pp. ISBN 0-7923-0487-X

M.A. Tsfasman and S.G. Vladuts: *Algebraic Geometric Codes*. 1991, 668 pp. ISBN 0-7923-0727-5

H.J. Voss: *Cycles and Bridges in Graphs*. 1991, 288 pp. ISBN 0-7923-0899-9

V.K. Kharchenko: *Automorphisms and Derivations of Associative Rings*. 1991, 386 pp. ISBN 0-7923-1382-8

A.Yu. Olshanskii: *Geometry of Defining Relations in Groups*. 1991, 513 pp. ISBN 0-7923-1394-1

F. Brackx and D. Constales: *Computer Algebra with LISP and REDUCE. An Introduction to Computer-Aided Pure Mathematics*. 1992, 286 pp. ISBN 0-7923-1441-7

N.M. Korobov: *Exponential Sums and their Applications*. 1992, 210 pp. ISBN 0-7923-1647-9

D.G. Skordev: *Computability in Combinatory Spaces. An Algebraic Generalization of Abstract First Order Computability*. 1992, 320 pp. ISBN 0-7923-1576-6

E. Goles and S. Martinez: *Statistical Physics, Automata Networks and Dynamical Systems*. 1992, 208 pp. ISBN 0-7923-1595-2

Other *Mathematics and Its Applications* titles of interest:

M.A. Frumkin: *Systolic Computations*. 1992, 320 pp. ISBN 0-7923-1708-4

J. Alajbegovic and J. Mockor: *Approximation Theorems in Commutative Algebra*. 1992, 330 pp. ISBN 0-7923-1948-6

I.A. Faradzev, A.A. Ivanov, M.M. Klin and A.J. Woldar: *Investigations in Algebraic Theory of Combinatorial Objects*. 1993, 516 pp. ISBN 0-7923-1927-3

I.E. Shparlinski: *Computational and Algorithmic Problems in Finite Fields*. 1992, 266 pp. ISBN 0-7923-2057-3

P. Feinsilver and R. Schott: *Algebraic Structures and Operator Calculus*. Vol. I. Representations and Probability Theory. 1993, 224 pp. ISBN 0-7923-2116-2

A.G. Pinus: *Boolean Constructions in Universal Algebras*. 1993, 350 pp. ISBN 0-7923-2117-0

V.V. Alexandrov and N.D. Gorsky: *Image Representation and Processing. A Recursive Approach*. 1993, 200 pp. ISBN 0-7923-2136-7

L.A. Bokut' and G.P. Kukin: *Algorithmic and Combinatorial Algebra*. 1994, 384 pp. ISBN 0-7923-2313-0

Y. Bahturin: *Basic Structures of Modern Algebra*. 1993, 419 pp. ISBN 0-7923-2459-5

R. Krichevsky: *Universal Compression and Retrieval*. 1994, 219 pp. ISBN 0-7923-2672-5

A. Elduque and H.C. Myung: *Mutations of Alternative Algebras*. 1994, 226 pp. ISBN 0-7923-2735-7

E. Goles and S. Martínez (eds.): *Cellular Automata, Dynamical Systems and Neural Networks*. 1994, 189 pp. ISBN 0-7923-2772-1

A.G. Kusraev and S.S. Kutateladze: *Nonstandard Methods of Analysis*. 1994, 444 pp. ISBN 0-7923-2892-2

P. Feinsilver and R. Schott: *Algebraic Structures and Operator Calculus*. Vol. II. Special Functions and Computer Science. 1994, 148 pp. ISBN 0-7923-2921-X

V.M. Kopytov and N. Ya. Medvedev: *The Theory of Lattice-Ordered Groups*. 1994, 400 pp. ISBN 0-7923-3169-9

H. Inassaridze: *Algebraic K-Theory*. 1995, 438 pp. ISBN 0-7923-3185-0

C. Mortensen: *Inconsistent Mathematics*. 1995, 155 pp. ISBN 0-7923-3186-9

R. Abłamowicz and P. Lounesto (eds.): *Clifford Algebras and Spinor Structures*. A Special Volume Dedicated to the Memory of Albert Crumeyrolle (1919–1992). 1995, 421 pp. ISBN 0-7923-3366-7

W. Bosma and A. van der Poorten (eds.), *Computational Algebra and Number Theory*. 1995, 336 pp. ISBN 0-7923-3501-5

Other *Mathematics and Its Applications* titles of interest:

A.L. Rosenberg: *Noncommutative Algebraic Geometry and Representations of Quantized Algebras*. 1995, 316 pp. ISBN 0-7923-3575-9

L. Yanpei: *Embeddability in Graphs*. 1995, 400 pp. ISBN 0-7923-3648-8

B.S. Stechkin and V.I. Baranov: *Extremal Combinatorial Problems and Their Applications*. 1995, 205 pp. ISBN 0-7923-3631-3

Y. Fong, H.E. Bell, W.-F. Ke, G. Mason and G. Pilz (eds.): *Near-Rings and Near-Fields*. 1995, 278 pp. ISBN 0-7923-3635-6

A. Facchini and C. Menini (eds.): *Abelian Groups and Modules*. (Proceedings of the Padova Conference, Padova, Italy, June 23–July 1, 1994). 1995, 537 pp. ISBN 0-7923-3756-5

D. Dikranjan and W. Tholen: *Categorical Structure of Closure Operators*. With Applications to Topology, Algebra and Discrete Mathematics. 1995, 376 pp. ISBN 0-7923-3772-7

A.D. Korshunov (ed.): *Discrete Analysis and Operations Research*. 1996, 351 pp. ISBN 0-7923-3866-9

P. Feinsilver and R. Schott: *Algebraic Structures and Operator Calculus*. Vol. III: Representations of Lie Groups. 1996, 238 pp. ISBN 0-7923-3834-0

M. Gasca and C.A. Micchelli (eds.): *Total Positivity and Its Applications*. 1996, 528 pp. ISBN 0-7923-3924-X

W.D. Wallis (ed.): *Computational and Constructive Design Theory*. 1996, 368 pp. ISBN 0-7923-4015-9

F. Cacace and G. Lamperti: *Advanced Relational Programming*. 1996, 410 pp. ISBN 0-7923-4081-7

N.M. Martin and S. Pollard: *Closure Spaces and Logic*. 1996, 248 pp. ISBN 0-7923-4110-4

A.D. Korshunov (ed.): *Operations Research and Discrete Analysis*. 1997, 340 pp. ISBN 0-7923-4334-4

W.D. Wallis: *One-Factorizations*. 1997, 256 pp. ISBN 0-7923-4323-9

G. Weaver: *Henkin–Keisler Models*. 1997, 266 pp. ISBN 0-7923-4366-2

V.N. Kolokoltsov and V.P. Maslov: *Idempotent Analysis and Its Applications*. 1997, 318 pp. ISBN 0-7923-4509-6

J.P. Ward: *Quaternions and Cayley Numbers*. Algebra and Applications. 1997, 250 pp. ISBN 0-7923-4513-4

E.S. Ljapin and A.E. Evseev: *The Theory of Partial Algebraic Operations*. 1997, 245 pp. ISBN 0-7923-4609-2

Other *Mathematics and Its Applications* titles of interest:

S. Ayupov, A. Rakhimov and S. Usmanov: *Jordan, Real and Lie Structures in Operator Algebras*. 1997, 235 pp. ISBN 0-7923-4684-X

MIX
Papier aus verantwortungsvollen Quellen
Paper from responsible sources
FSC® C105338

If you have any concerns about our products,
you can contact us on
ProductSafety@springernature.com

In case Publisher is established outside the EU,
the EU authorized representative is:
Springer Nature Customer Service Center GmbH
Europaplatz 3, 69115 Heidelberg, Germany

Printed by Libri Plureos GmbH
in Hamburg, Germany